AF839688

Profiling Water Quality Parameters: From Source Water to the Household Tap

The mission of the Awwa Research Foundation (AwwaRF) is to advance the science of water to improve the quality of life. Funded primarily through annual subscription payments from over 1,000 utilities, consulting firms, and manufacturers in North America and abroad, AwwaRF sponsors research on all aspects of drinking water, including supply and resources, treatment, monitoring and analysis, distribution, management, and health effects.

From its headquarters in Denver, Colorado, the AwwaRF staff directs and supports the efforts of over 700 volunteers, who are the heart of the research program. These volunteers, serving on various boards and committees, use their expertise to select and monitor research studies to benefit the entire drinking water community.

Research findings are disseminated through a number of technology transfer activities, including research reports, conferences, videotape summaries, and periodicals.

Profiling Water Quality Parameters: From Source Water to the Household Tap

Prepared by:
Mark W. LeChevallier, **Mohammad Karim**, **Ramon Aboytes**, **Richard Gullick**, and **Janice Weihe**
American Water, Voorhees, NJ 08043

Brock Earnhardt, **Joel Mohr**, **Judy Starcevich**, and **Jim Case**
Iowa-American Water Company, Davenport, IA 52805

Jeffrey S. Rosen and **Jose Sobrinho**
Technology Planning & Management Consultants, Scituate, MA 02066

Jennifer L. Clancy and **Randi M. McCuin**
Clancy Environmental Consultants, St. Albans, VT 05478

and

James E. Funk and **Don J. Wood**
University of Kentucky, Lexington, KY 40506-0108

Sponsored by:
Microbial/Disinfection By-Products Research Council

Jointly funded by:
Awwa Research Foundation
6666 West Quincy Avenue, Denver, CO 80235-3098

and

U.S. Environmental Protection Agency
1200 Pennsylvania Avenue, N.W., Washington, DC 20460

Published by:

Awwa Research Foundation

IWA Publishing

DISCLAIMER

This study was funded by the Awwa Research Foundation (AwwaRF). AwwaRF assumes no responsibility for the content of the research study reported in this publication or for the opinions or statements of fact expressed in the report. The mention of trade names for commercial products does not represent or imply the approval or endorsement of AwwaRF. This report is presented solely for informational purposes.

Published by IWA Publishing, Alliance House, 12 Caxton Street, London SW1H 0QS, UK

Telephone: +44 (0) 20 7654 5500; Fax: +44 (0) 20 7654 5555; Email: publications@iwap.co.uk;
Web: **www.iwapublishing.com**

AwwaRF report number 90986F. Originally published by AwwaRF for its subscribers in 2003.
IWA Publishing version published 2004.

ISBN 1 84339 8788

CONTENTS

TABLES

FIGURES

FOREWORD

The Awwa Research Foundation is a nonprofit corporation that is dedicated to the implementation of a research effort to help utilities respond to regulatory requirements and traditional high-priority concerns of the industry. The research agenda is developed through a process of consultation with subscribers and drinking water professionals. Under the umbrella of a Strategic Research Plan, the Research Advisory Council prioritizes the suggested projects based upon current and future needs, applicability, and past work; the recommendations are forwarded to the Board of Trustees for final selection. The foundation also sponsors research projects through the unsolicited proposal process; the Collaborative Research, Research Applications, and Tailored Collaboration programs; and various joint research efforts with organizations such as the U.S. Environmental Protection Agency, the U.S. Bureau of Reclamation, and the Association of California Water Agencies.

This publication is a result of one of these sponsored studies, and it is hoped that its findings will be applied in communities throughout the world. The following report serves not only as a means of communicating the results of the water industry's centralized research program but also as a tool to enlist the further support of the nonmember utilities and individuals.

Projects are managed closely from their inception to the final report by the Foundation's staff and large cadre of volunteers who willingly contribute their time and expertise. The Foundation serves a planning and management function and awards contracts to other institutions such as water utilities, universities, and engineering firms. The funding for this research effort comes primarily from the Subscription Program, through which water utilities subscribe to the research program and make an annual payment proportionate to the volume of water they deliver and consultants and manufacturers subscribe based on their annual billings. The program offers a cost-effective and fair method for funding research in the public interest.

A broad spectrum of water supply issues is addressed by the Foundation's research agenda: resources, treatment and operations, distribution and storage, water quality and analysis, toxicology, economics, and management. The ultimate purpose of coordinated effort is to assist water suppliers to provide the highest possible quality of water economically and reliably. The true benefits are realized when the results are implemented at the utility level. The Foundation's trustees are pleased to offer this publication as a contribution toward that end.

In response to the 1996 amendments to the Safe Drinking Water Act requirements, the U.S. Centers for Disease Control and Prevention (CDC) and the U.S. Environmental Protection Agency (EPA) formulated a series of epidemiology studies to determine the occurrence of waterborne disease in the U.S. One of these epidemiology studies was conducted in Davenport, Iowa. To supplement the epidemiology study, the Awwa Research Foundation funded a study to examine water quality parameters, from the source water, through the treatment process, distribution system, to the household tap. It was extremely important to thoroughly characterize any changes in water quality so that the results of the epidemiology study could be placed within the proper operational and regulatory context. The project used some of the latest advances in microbiological, monitoring, and modeling techniques. The project is an important advance in the science of water analysis to ensure that the best information is obtained to make informed decisions to benefit the health and safety of drinking water consumers.

Edmund G. Archuleta, P.E.
Chair, Board of Trustees
Awwa Research Foundation

James F. Manwaring, P.E.
Executive Director
Awwa Research Foundation

PREFACE

The Microbial/Disinfection By-Product Council was established in 1995 as a vehicle for the selection and funding of research to provide scientific information in the areas of health effects, exposure assessment, risk assessment, and prevention and control of contamination by microbes and disinfection by-products in drinking water. The council is composed of representatives designated by the U.S. Environmental Protection Agency (USEPA), the Awwa Research Foundation (AwwaRF) Board of Trustees, the Association of State Drinking Water Administrators, the National Resources Defense Council, the National Environmental Health Association, or their designees. Sources of funding for this research include the USEPA and AwwaRF, along with other interested parties. The council disburses these funds for research deemed to be of the highest urgency and importance in resolving critical research issues in drinking water.

ACKNOWLEDGMENTS

The authors would like to thank the Awwa Research Foundation and the United States Environmental Protection Agency for funding and support. This project was also funded by the utility subsidiaries of the American Water. The authors wish to thank Ms. Linda Reekie, project manager for the Awwa Research Foundation, Denver, Colorado and the following members of the Project Advisory Committee for their guidance and suggestions: Donald Reasoner, USEPA, Cincinnati, Ohio; Anita Highsmith, Highsmith Environmental Consultants, Inc., Atlanta, Georgia; Susan Shaw, USEPA, Washington, DC; and Nira Yamachika, Orange County Water District, Fountain Valley Calif. The contributions of the following personnel from American Water's Quality Control and Research Laboratory in Belleville, Illinois are acknowledged: Stacey Dulaney, Wyndi Klement, and Michael Spinner. The investigators wish to thank the many associates of the Iowa-American Water Company, Davenport, Iowa, for their outstanding contributions and commitment to the project.

EXECUTIVE SUMMARY

STUDY OBJECTIVES

In response to the 1996 amendments to the Safe Drinking Water Act requirements, the U.S. Centers for Disease Control and Prevention (CDC) and the U.S. Environmental Protection Agency (EPA) formulated a series of epidemiology studies to determine the occurrence of waterborne disease in the U.S. Dr. Jack Colford of the University of California at Berkeley School of Public Health conducted one of these epidemiology studies in collaboration with the Iowa-American Water Company in Davenport, Iowa. The study is a randomized, triple-blinded, placebo-controlled, cross-over intervention study of 456 households using both active and inactive UV and micro-filtration devices. The results of the epidemiology study will be reported elsewhere.

The Awwa Research Foundation funded American Water to conduct a water quality study in the Davenport area in parallel to the epidemiology study. The primary objective of the project was to characterize raw water, treatment plant, distribution, and household tap water quality with respect to microbes, water quality indictors, and physicochemical parameters so as to provide a context for interpreting the public health data generated by the epidemiology study.

STUDY LOCATION

The Iowa American Water Company, Davenport district, cooperated fully with the epidemiology study. The Davenport area was selected because it best satisfied all of the following criteria:

- The entire community had to receive its drinking water from one surface water (river) source;
- The source of supply should be microbially challenged;
- The source water had to be treated at one water treatment plant;
- The water had to be treated by conventional drinking water treatment methods;
- The community had to be large enough (about 100,000 people) to do a study on 400 households;
- The community also had to be fairly representative of a typical American metropolitan area.

It should be emphasized that Davenport was not selected because of any problem with the water quality or the water treatment process. In fact, available data showed that drinking water in the Davenport area met or was better than all federal and State guidelines or safety standards.

PROJECT APPROACH

Detailed analyses were conducted on raw, treated plant effluent, distribution system, and household tap and treatment device water samples. Analyses included *Cryptosporidium* oocysts by both USEPA Method 1623 and infectious organisms by the cell culture/polymerase chain reaction method (CC-PCR), *Giardia* cysts by Method 1623, culturable viruses, total and fecal coliforms, *Clostridium perfringens* spores, *Bacillus subtilis* spores, male-specific and somatic bacteriophage, and HPC bacteria. Monitoring of the treatment plant included a large number of process variables as well as microbial indicators. Data on distribution system leaks, electronic pressure monitoring for hydraulic surges, and hydraulic modeling helped to supplement the monitoring program.

RAW WATER DATA

Analysis of the raw water data showed the presence of fecal pathogens in the Mississippi River. Using Method 1623, *Giardia* cysts were detected in 31.1% of the weekly samples with an average of 0.093 cysts/L, and *Cryptosporidium* oocysts were detected in 13.5% of the samples with an average of 0.019 oocysts/L. Infectious *Cryptosporidium* were observed in 2.8% of the samples using the CC-PCR method. Culturable viruses were detected in 57.9% of the monthly raw water samples, with an average 6.72 MPN/100 L. The occurrence and concentration of these pathogens increased during the colder months and were generally absent during the summer. The Mississippi River experienced a flood between April 14 and May 23, 2001, and increased concentrations were observed during this period for *Giardia*, male-specific coliphage, total and fecal coliforms. Comparison of microbial results with previous databases showed that the source water quality parameters measured during the study were fairly representative of the Mississippi River at Davenport, Iowa.

TREATMENT PLANT DATA

Extensive monitoring of the Iowa-American Water Company, East River Station was conducted to document the plant performance for control of microbial pathogens and the production of consistent water quality. All the data indicated that the plant met or surpassed all drinking water criteria. No organisms of public health significance were detected in the treated water. The treatment plant was uniformly negative for all measurements of infectious *Cryptosporidium*, culturable enteric viruses, total coliforms, somatic coliphage, and *Clostridium perfringens* spores. Measurement of plant performance ranged from 1.5 to >6.36 log removal of microbes or particulate material, with the observed level of plant performance generally limited by the raw water concentration. The average plant effluent turbidity was 0.05 NTU, and for the 20 individual filters the average effluent turbidity values ranged from 0.034 to 0.055 NTU. Of the 10,991 daily turbidity values recorded, only four exceeded 0.15 NTU and none was greater that 0.45 NTU. The level of disinfection averaged 7 times greater than the minimum (0.5 log) required for *Giardia* inactivation.

Despite raw water flood conditions between April 14 and May 23, 2001 finished water quality remained excellent with individual filter turbidities and combined filter effluent turbidities consistently less than 0.10 NTU. Filtered water particle counts were low, averaging less than 18 particles/mL, and free chlorine CxT values averaging 3.5 times higher than the minimum required level of *Giardia* inactivation.

DISTRIBUTION SYSTEM WATER QUALITY DATA

The Iowa-American Water Company Davenport district distribution system includes more than 541 miles of mains within seven pressure zones. By all conventional measures of water quality, the drinking water met or surpassed all standards. Over 99.9% of the distribution system samples were free of total coliform bacteria, and 100% were free of *E. coli*. The average distribution system total chlorine residual was 2.08 mg/L and the minimum total chlorine level detected (based on 2,471 samples) was 0.59 mg/L. The average HPC bacterial level was 230 cfu/mL (measured using R2A agar at 20°C for 7 days). Male-specific coliphage were detected in 5.6% of the distribution samples (average concentration of 0.1 pfu/100 mL), and were speculated to be associated with main breaks during the winter months. However, the association between main breaks and the detection of coliphage was not particularly strong and additional research will be necessary to establish further relationships between these parameters.

PRESSURE MONITORING AND MODELING

Extensive pressure monitoring of the Davenport distribution system using seven electronic data loggers collected a combined 3,190 days of pressure monitoring totaling 2.756×10^8 data points and found only five occasions when distribution system pressures were less than 20 psi. No negative pressures were observed. Although the cause of these low-pressure events is not fully known, it is thought that most events were caused by pump shutdowns. Distribution system modeling suggested that significant low pressures ([0 psi) were possible when the East River pumping station lost power. The model simulation suggested that protection against low pressures is possible by installing surge tanks downstream of the pumping stations. Modeling results suggested that air/vacuum valves did not offer reliable protection against low pressures in the system.

HOUSEHOLD TAP AND TREATMENT DEVICE DATA

The objective of the household monitoring was to characterize the water quality of the treatment devices (active and inactive UV and micro-filtration point-of-use devices) and kitchen taps, and to compare the data to results from the treatment plant and distribution system to determine if significant changes had occurred in the microbial quality of water. Overall, the values for pH, turbidity, temperature, and total chlorine residual were not substantially different between the tap and device samples. However, the detection of total coliform bacteria was 6.5 times more frequent, and levels were 20 times higher, in the treatment device samples than in the kitchen tap samples. Analysis of several treatment devices detected total coliform bacteria within the diverter valve and in the UV treatment unit, suggesting that the bacteria were growing within the devices. HPC bacterial levels were four times higher in the devices compared to the taps, also supporting the regrowth hypothesis. Total coliform bacterial species identified in the units and tap water samples were those typically associated with regrowth in distribution system biofilms. The detection of one *E. coli* positive device sample suggests that the coliforms may have originated from contamination of the faucets by household activities (food preparation, dish washing, etc.) in the kitchen area.

CONCLUSIONS

The extensive monitoring of the Davenport system failed to find significant indicators of microbial risk. The drinking water in the Davenport system met or surpassed all measures of water quality. Despite using a "microbially challenged" source water, the operators of the East River Station met the challenge by consistently producing a high quality product. The resiliency of the treatment process was evident during the Spring 2001 flood when, despite degraded raw water quality, finished drinking water quality remained outstanding.

The study did identify some areas for future consideration and research, including the relationship between the occurrence of male-specific coliphage and distribution system main breaks, the potential for negative pressure transients due to power outages, and the need for more comprehensive distribution system pressure monitoring. However, there was no clear link between these events and the degradation of drinking water quality.

The study also provided direction to utilities to resolve some of these issues. Installation of surge control tanks down stream of the high service pumps in the Davenport System was predicted to control the potential for negative pressure transients due to power outages. Distribution systems models were shown to be useful to guide the placement of electronic pressure monitors in areas that may be most vulnerable to low pressure events. Electronic reporting of low-pressure events could better alert treatment plant operators to events within the treatment process that may trigger water quality problems within the distribution system. Coliphage monitoring, especially in chloraminated systems, may be a sensitive indicator of contamination due to cold-weather main breaks.

Finally, the microbial benefits of point-of-use treatment devices remains in doubt. The levels of HPC and total coliform bacteria found within the device raise concerns that units may be generators of microbial risk. Additional studies are needed to examine the pathways and the types of microbes (*Legionella*, *Mycobacterium*, enteric pathogens, etc.) that may contaminate point-of-use devices in the kitchen environment. Studies of household cross connections should also prove useful in addressing the total microbial risk that can be transmitted through water.

CHAPTER 1
PROJECT BACKGROUND AND LITERATURE REVIEW

OVERVIEW OF WATERBORNE DISEASE IN THE UNITED STATES

Bennett et al. (1987) estimated the incidence of waterborne disease in the United States was 940,000 cases per year, and resulted in 900 deaths. Although the purpose of the study was to rank the importance of various disease categories (water ranked next to the last, above zoonotic diseases) and to define the opportunities for prevention, the study has been criticized as little more than an educated guess. Morris and Levin (1995) gave waterborne infectious disease estimates of 560,000 cases of moderate or severe illness, 7.1 million mild infections, and 1,200 deaths. However, because of large uncertainties regarding the level of waterborne disease in the United States, the U.S. Congress amended the Safe Drinking Water Act in 1996 to require the U.S. Centers for Disease Control and Prevention (CDC) and the U.S. Environmental Protection Agency (EPA) to conduct studies to determine the occurrence of waterborne disease. This requirement was needed because of problems with proper diagnosis and underreporting of cases of waterborne disease by medical practitioners, inadequate waterborne disease surveillance programs and chemical exposure registries, and the lack of financial resources and epidemiologists to fully investigate waterborne outbreaks.

The CDC and the USEPA maintain a collaborative system for collecting and reporting waterborne disease outbreak data for both microbial and chemical agents. However, detection and investigation of waterborne outbreaks is primarily the responsibility of the local, State, and territorial public health departments with voluntary reporting to the CDC (USEPA 2001a). From 1971 to 1998, 619 drinking waterborne disease outbreaks due to a microbiological, chemical, or unknown etiology were reported in the United States (Barwick et al. 2000). Since the mid-1980s, the number of community waterborne outbreaks has declined (Figure 1.1). The reason for the decrease is largely attributed to the promulgation of more stringent drinking water standards, including the Surface Water Treatment Rule, the Total Coliform Rule, and other regulations (USEPA 1989a,b, 1990, 1991, 1996a, 1998c). In addition, many water utilities have made voluntary improvements, such as the Partnership For Safe Water program to reduce the risk of waterborne cryptosporidiosis. The Partnership program entails a comprehensive evaluation of treatment practices with a focus on achieving filtered drinking water turbidities less than 0.1 nephelometric turbidity units (NTU).

The agents responsible for these outbreaks were predominantly undefined, microbial (parasitic, bacterial, or viral), or chemical (Figure 1.2). The large number of parasitic outbreaks in the early 1980s was mostly caused by *Giardia*, and was greatly reduced by the implementation of the Surface Water Treatment Rule in 1989 (Barwick et al. 2000). Relatively few outbreaks due to viruses have been reported, in part because of the difficulty of the detection methodologies for these organisms. To better address the occurrence of microbial pathogens in drinking water, the USEPA has issued a Contaminant Candidate List that includes 11 microbes for methods development and potential future regulation (USEPA 1998d).

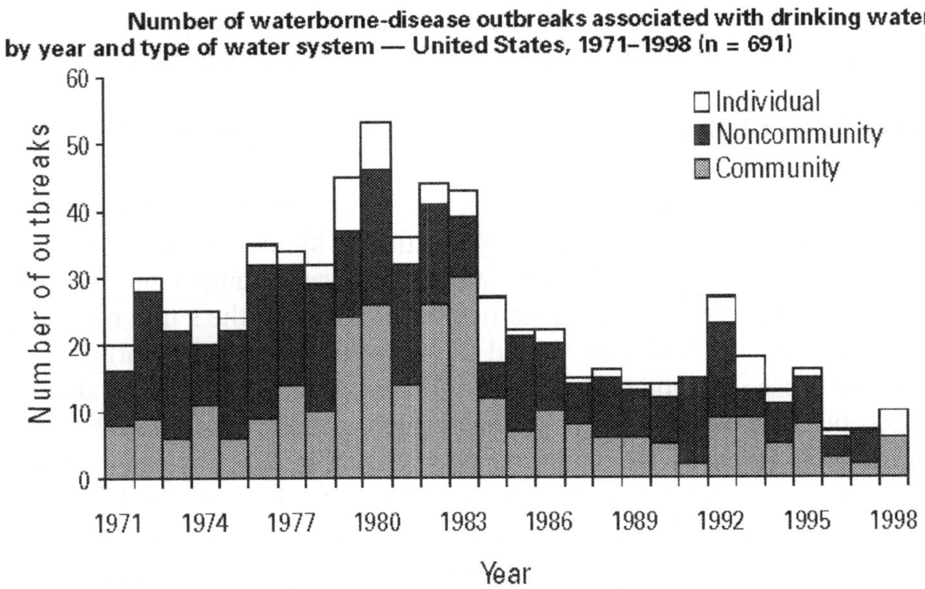

Figure 1.1. Number of drinking water disease outbreaks in the United States, 1971-1998. Individual, private or individual water systems (9% of US population or 24 million users); Community, systems that serve >25 users year round (91% of US population or 243 million users); Noncommunity, systems that serve <25 users and transient water systems such as restaurants, highway rest areas, parks (millions of users yearly). Adapted from Barwick et al. 2000.

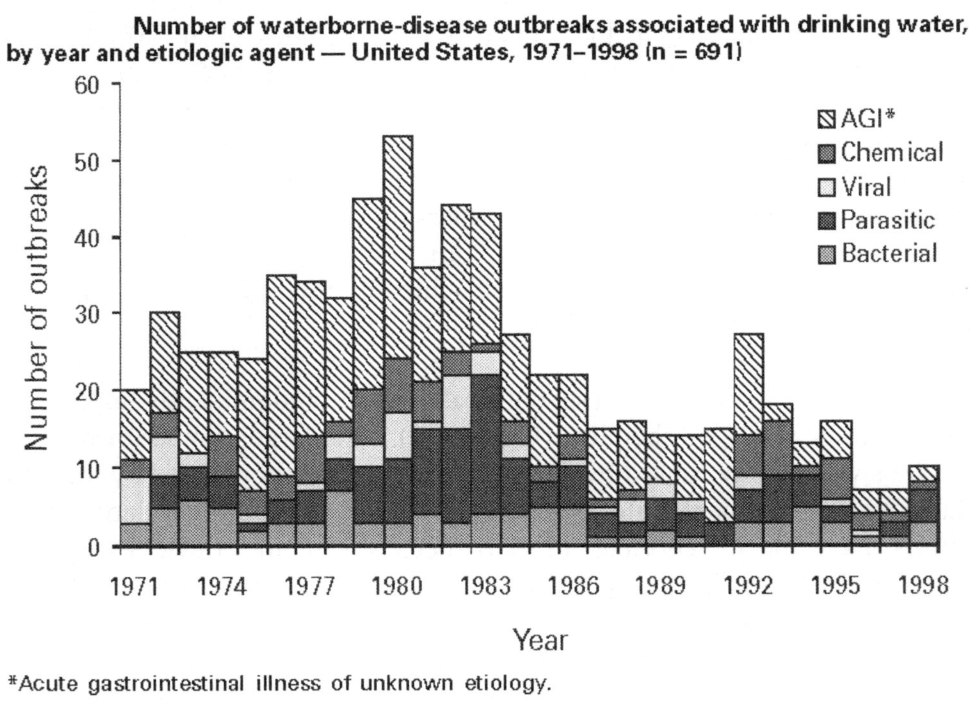

Figure 1.2. Agents responsible for waterborne outbreaks. Adapted from Barwick et al. 2000.

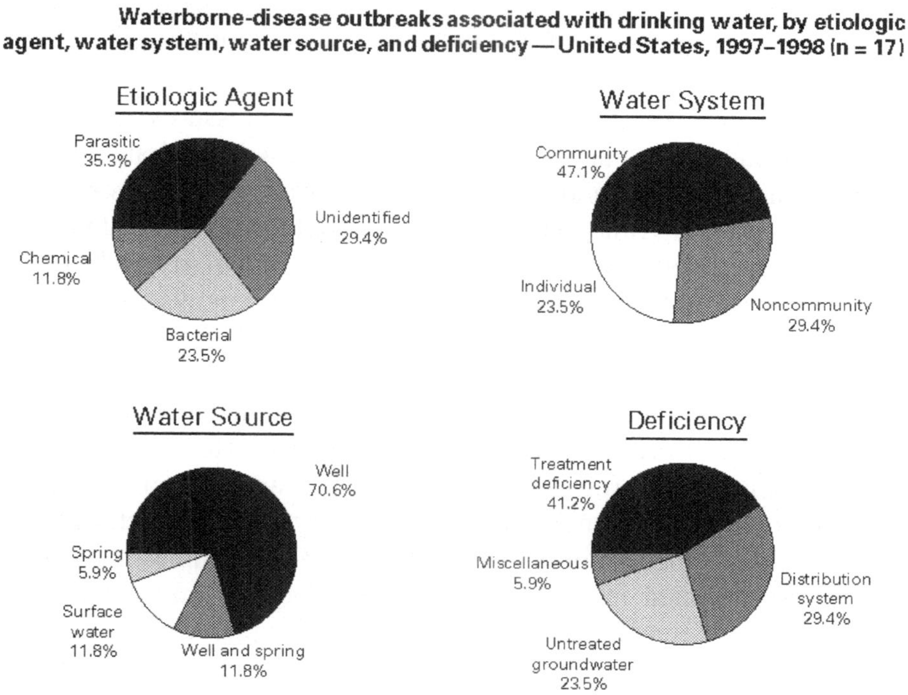

Figure 1.3. Waterborne outbreaks by etiological agent, water system, water source, and deficiency, 1997-1998. Adapted from Barwick et al. 2000.

In recent years waterborne outbreaks have increasingly involved groundwater supplies or distribution system deficiencies (Figure 1.3). The USEPA has identified these areas for regulatory consideration. One area of growing attention is the potential for distribution system contamination through cross connections and transient low distribution system pressures that could lead to intrusion of contaminants into the distribution system (Kirmeyer et al. 2001, LeChevallier et al. 2003).

PRIOR EPIDEMIOLOGY STUDIES

Payment et al. conducted two epidemiology studies (Payment et al, 1991; Payment et al, 1997), each suggesting that the distribution system was at least partially responsible for increased levels of gastrointestinal illnesses. The studies examined the health of people who drank tap water and compared the group to people receiving water treated by reverse osmosis to determine which group had higher levels of gastrointestinal illness. In the 1991 study, reverse osmosis units were installed in 299 households (1,206 persons) and another 307 households (1,202 persons) were followed as controls with no device installed. Both groups were monitored for a 15-month period, after which the authors concluded that 35% of the self-reported gastrointestinal illnesses were attributed to tap water consumption. The 1997 study included groups receiving regular tap water, tap water from a continuously purged tap, bottled plant effluent water, and purified (by reverse osmosis) plant effluent water. This second study attributed 14% to 40% of the gastrointestinal illness to the consumption of tap water. In both studies, the quality of tap water met Canadian guidelines (Payment et al. 1991, 1993). Analysis of Payment's data shows that people who lived in zones far away from the treatment plant had the highest risk of gastroenteritis. Transient pressure modeling (Kirmeyer et al. 2001) found that the distribution system studied by Payment was extremely prone to negative pressures, with more than 90 percent of the nodes within the system drawing negative pressures under certain modeling scenarios (e.g., power outages). The system, located in the Montreal area, reported many pipe breaks, particularly during the Fall and Winter when temperature changes placed added stresses on the distribution system. Although the system employed state-of-the-art treatment, the distribution network maintained low disinfectant residuals, particularly at the ends of the system. Low disinfectant residuals and a vulnerability of the distribution system to pressure transients could account for the viral-like etiology of the illnesses observed.

A double-blinded, randomized, trial was recently completed in Melbourne, Australia, to determine the contribution of drinking water to gastroenteritis (Hellard et al. 2001). Melbourne draws its drinking water from a protected forest watershed and has an unfiltered surface water supply using only free chlorine treatment. Free chlorine levels in the distribution system ranged from 0 to 0.94 mg/L, with a median of 0.05 mg/L, and 90% of samples had < 0.20 mg/L. Total coliform bacteria were detected in 18.9% of 1,167 routine 100-mL water samples, but fecal coliform bacteria were not detected. Distribution system samples were positive for *Aeromonas* spp. (50% of 68 weekly samples), *Campylobacter* (1 occasion) and *Giardia* (2 viable samples by reverse transcriptase-polymerase chain reaction). Six hundred families were randomly assigned to receive either a real or placebo water treatment unit installed under the kitchen sink. Functional units were designed to remove viruses, bacteria, and protozoa using microfiltration and ultraviolet light treatment. Study participants completed a weekly health diary reporting gastrointestinal symptoms during the 68-week observation period. The study found that the water was not a source of measurable gastrointestinal disease (the ratio of illness rates between the group drinking treated water compared to the normal tap water was 0.99, with a 95% confidence interval of 0.85–1.15; $p = 0.85$). Analysis of 795 fecal specimens from participants with gastroenteritis did not reveal any difference in pathogen detection rates between the two groups.

As a preliminary trial to the current epidemiology study, a randomized, triple-blinded, home drinking water intervention trial of 77 households was conducted for four months (Colford et al, 2002). The drinking water was treated using an under the kitchen sink device that incorporated ultraviolet light and microfiltration. Although the purpose of the trial was to evaluate the "blinding" of the study (e.g., could the participating households detect the active and identical-looking placebo devices), analysis of the data showed that the incidence rate ratio of disease (adjusted for clustering) was 1.32, with a 95% confidence interval of 0.75 to 2.33. Given the small study size, the higher rate of gastrointestinal disease among the placebo group was not statistically significant, the authors concluded, however, that the relative rates of illnesses were similar to those observed by Payment et al. (1991, 1997).

A cross-sectional survey of 226 HIV-infected individuals in San Francisco (an unfiltered drinking water system) found that 53% consumed regular tap water, and that the untreated tap water was not associated with increased illnesses (Eisenberg et al. 2002). The researchers found that the immune status of the individuals (measured by CD4 blood counts), and the level of medication were significantly associated with gastrointestinal illnesses.

STUDY OBJECTIVES

In response to the 1996 amendments to the Safe Drinking Water Act requirements, the U.S. Centers for Disease Control and Prevention (CDC) and the U.S. Environmental Protection Agency (EPA) formulated a series of epidemiology studies to determine the occurrence of waterborne disease in the U.S. Dr. Jack Colford of the University of California at Berkeley School of Public Health conducted one of these epidemiology studies in collaboration with the Iowa-American Water Company in Davenport, Iowa.

The American Water Works Association Research Foundation and the USEPA funded American Water to conduct a water quality study in the Davenport area in parallel to the epidemiology study. The primary objective of the project was to characterize raw water, treatment plant, distribution, and household tap water quality with respect to microbes, water quality indictors, and physicochemical parameters so as to provide a context for interpreting the public health data generated by the epidemiology study. Because the project built on previous AwwaRF studies of pathogen occurrence in sources waters (LeChevallier et al. 2002), treatment plant performance (DiGiovanni et al. 1999), and distribution system water quality (Kirmeyer et al. 2001), it was of interest to compare the data generated by the current study to previous work, and to use the information to target areas for future research.

Raw water monitoring built on previous studies that actively monitored the Mississippi River water (LeChevallier et al. 2002). Analyses included *Cryptosporidium* oocysts by both USEPA Method 1623 and the cell culture/polymerase chain reaction method (CC-PCR) (DiGiovanni et al. 1999), *Giardia* cysts by Method 1623, culturable viruses, total and fecal coliforms, *Clostridium perfringens* spores, *Bacillus subtilis* spores, male-specific and somatic bacteriophage, algae, alkalinity, hardness, pH, turbidity, temperature, river stage, flow rate, and precipitation.

Monitoring of the treatment plant included a large number of process variables (chemical feed rates, flow rate, loss of head, filter run time, filter backwashes, turbidity data for raw, settled, individual filter effluents, and combined plant effluent, particle counts for settled and combined plant effluent, disinfectant concentration multiplied by contact time (CxT) calculations, and other parameters) as well as microbial indicators (coliforms, HPC bacteria, *Clostridium perfringens* spores, *Bacillus subtilis* spores, and male-specific and somatic bacteriophage) and infectious pathogens (*Cryptosporidium* oocysts and culturable viruses).

Distribution system monitoring built upon the State-approved coliform monitoring plan and added measurements of temperature, pH, phosphate, free and total chlorine, HPC bacteria, and male-specific and somatic bacteriophage analyses. Supplementing the monitoring plan was the collection of data on distribution system leaks, electronic pressure monitoring to detect hydraulic surges that could result in intrusion into the distribution system, and hydraulic modeling to determine the vulnerability of the system to intrusion under various modeling scenarios.

The household monitoring task documented microbial water quality at household taps and from the treatment devices so that comparisons could be made between the treatment devices, taps, distribution system and plant effluent data. Parameters measured included: total chlorine, pH temperature, total coliforms, HPC bacteria, and turbidity.

The success of the project was dependent upon the expertise of the project team, including: Clancy Environmental Consultants, Inc. (CEC) who performed the *Giardia* and *Cryptosporidium* analyses using Method 1623; Technology Planning Management Corporation (TPMC) who performed the data management and statistical analysis; the University of Kentucky who performed the hydraulic and surge modeling; Iowa American Water Company who provided access to the system and provided data and expertise in interpreting the results; and the research staff of American Water who performed analyses of infectious pathogens (*Cryptosporidium* and culturable viruses), microbial indicators, and specialized pressure monitoring.

CHAPTER 2
THE WATER EVALUATION TRIAL

STUDY GOALS

In 1996, amendments to the Safe Drinking Water Act required the U.S. Centers for Disease Control and Prevention (CDC) and the U.S. Environmental Protection Agency (EPA) to conduct studies to determine the occurrence of waterborne disease for at least five major communities or public water systems and prepare a report on the findings (Pontius 1997). The primary objective of the Water Evaluation Trial was to determine the incidence of gastrointestinal illness associated with consumption of drinking water meeting all federal and state treatment guidelines. The study was conducted by a team of epidemiologists lead by Dr. Jack Colford of the University of California at Berkeley School of Public Health.

To offset some of the criticisms of previous epidemiological studies, the Water Evaluation Trial used a randomized, triple-blinded, placebo-controlled, crossover, intervention design (Colford et al, 2002). The intervention tested household-level treatment of drinking water using a kitchen countertop device that treated tap water with ultraviolet light and microfiltration (Figure 2.1). Participating households were randomly assigned to two different groups. One group received the active device and the other received an identical-looking placebo (inactive) device. Half way through the study, "cross-over" occurred with the active devices replaced with inactive devices and the inactive devices replaced with active devices. The participants, the study staff, and the data analysis team were "blinded" (unaware of) to which group each household was assigned throughout the study. Blinding was necessary to reduce the likelihood of bias in reporting of disease and in the interpretation of the study results.

Secondary goals of the study were to evaluate specific pathogen(s) responsible for waterborne illness, to estimate the risk (if any) of waterborne disease to specific subgroups such as children or the elderly, to provide estimates of gastrointestinal illnesses in the general population, to evaluate the health benefit (if any) and the impact of point of use devices on drinking water quality, and to assess drinking water consumption patterns in the general population.

STUDY LOCATION

The Iowa American Water Company, Davenport district cooperated fully with the epidemiology study. Iowa American Water Company has an extensive water quality monitoring program that includes monitoring of raw water at the treatment plant intake pipe, the water leaving the treatment plant, and water throughout the distribution system. The current water quality study enhanced the existing monitoring program by collecting more frequent samples and analyzing the samples for specialized pathogens and indicator organisms. The Davenport area was selected because it best satisfied all of the following criteria:

- The entire community had to receive its drinking water from one surface water (river) source;
- The source of supply should be microbially challenged;

- The source water had to be treated at one water treatment plant;
- The water had to be treated by conventional drinking water treatment methods;
- The community had to be large enough (about 100,000 people) to do a study on 400 households;
- The community also had to be fairly representative of a typical American metropolitan area.

It should be emphasized that Davenport was not selected because of any problem with the water quality or the water treatment process. In fact, available historical data showed that drinking water in the Davenport area met or was better than all federal guidelines and safety standards.

Figure 2.1. Study treatment unit opened to show the microfilter (rounded vessel) and ultraviolet light (square box). During the study the units were sealed to prevent tampering.

BASIC STUDY DESIGN

The study began in November 2000, and was completed in June 2002. Households in the Davenport area were recruited to participate in the study through a letter and a brochure mailed to Iowa-American customers. Interested households that met enrollment criteria and consented to participate were enrolled in the study and water treatment devices were installed at the kitchen faucet in 456 households. Participants were instructed to drink water from the device and to take bottles of the water with them when they were away from home. Household participants in both groups completed questionnaires and health diaries for one year, reporting symptoms of gastrointestinal and respiratory illness, including diarrhea, abdominal cramps, nausea, fever, cough, or vomiting. A subset of 156 (34.2%) households from both groups was recruited to provide stool specimens during a gastrointestinal illness, and blood specimens at several points during the study. These samples were tested for common waterborne pathogens.

The crossover process (switching from active to placebo or *vice versa*) began in June 2001, and all households completed the crossover in December 2001. A total of 412 (90% of the original 456) households were crossed over. The final device was removed from the last remaining household in June 2002. A total of 387 households (85% of the original 456 households) completed the entire year of follow up.

Of the 156 households providing clinical samples, water samples were collected from 107 households and examined for various water quality parameters. Samples were collected from both the kitchen faucet tap and the treatment devices at the time of installation, upon crossover, and when the units were removed at the end of the study. To ensure anonymity of the participants, only the epidemiology team had direct contact with the households. This meant that members of the epidemiology team had to collect the samples and deliver them to the Iowa-American laboratory for analysis.

CROSS-CONNECTION SURVEY

A team from the University from Southern California (USC) received a follow-up grant to examine 200 of participant households for cross-connections. The UC Berkeley team will use the information to determine if the plumbing condition in a house could be associated with the occurrence of gastrointestinal illnesses. A local expert in cross-connections control was contracted to perform the inspections. The survey was completed at the end of 2002.

CHAPTER 3
SITE CHARACTERIZATION

WATERSHED CHARACTERIZATION

Background information and water quality data were collected and compiled for Upper Mississippi River (UMR) watershed upstream from the Davenport, Iowa intake. A variety of national and regional environmental databases: the US Environmental Protection Agency (USEPA) BASINS (USEPA 1998b), USEPA Surf your Watershed (USEPA 1998a), United States Geological Survey (USGS 1996), and USEPA ENVIROFACTS were most consulted for acquisition of data pertaining to the selected site. At the regional level, many state agencies provided a variety of information and data. Iowa American Water Company also provided other background information as well as raw water quality data.

Watershed Description

The 2,348-mile long Mississippi River flows from the headwaters at Lake Itasca in northern Minnesota to its mouth in the Gulf of Mexico in Louisiana (Figure 3.1). The Mississippi River drains 41 percent of the conterminous United States with approximately 30 percent of the United State's population. The principal tributaries of the Mississippi River include the Minnesota, St. Croix, Wisconsin, Illinois, Kaskaskia, Missouri, Ohio, Arkansas, and the Yazoo Rivers.

The Ohio confluence at Cairo, Illinois divides the Mississippi River in two general hydrologic regions: the upper (1,300 miles) and lower (1,000 miles) regions. The commercially navigable portion of the Upper Mississippi River between Minneapolis, MN and St. Louis, MO is characterized by twenty-nine locks and dams (pools) that control the depth and, to a lesser extent, the flow of the river (UMRBA, 1997). Davenport, IA (Lower UMR, pool 15) is located in the USGS hydrologic unit catalog (HUC) 07080101, Copperas-Duck (Figure 3.2). This area defines the source of the USGS data for the Davenport watershed.

Watershed Management/Land Use

The Upper Mississippi River is a significant commercial transportation, wildlife, recreational, and water supply resource. The Upper Mississippi River is the only inland federal commercial navigation route, which also serves as a major national wildlife refuge. In addition, the Upper Mississippi River supplies water to 70 cities and towns (USGS 1996). Davenport, IA withdraws water from the Mississippi River to supply customers in its vicinities.

Land uses in the Upper Mississippi River basin include urban, industrial, and agricultural (Barber et al., 1998). Rural land is spread over most of the watershed and is comprised of both crop use (corn and soybean) and livestock (cattle and pigs). The river receives direct effluent discharges from a variety of industries and municipalities (combined sewer overflows and sewerage treatment plants). According to information from the management of the Iowa American Water Company, Davenport, there are more than ten wastewater treatment plants discharging upstream of their intake.

Approximate percentages of the Davenport watershed study area dedicated to specific land uses shown in Table 3.1 were calculated by the Sample Grid Estimation Method from USGS land use maps. The percentage of the watershed with impervious surfaces (25%) was based on the amount of land that was classified as residential, commercial, or industrial.

Water Quality

The Mississippi River raw water at Davenport, IA is sampled for protozoa, total and fecal coliforms, and turbidity. Samples are collected near the water plant intake by the utility personnel. Historical descriptive statistics for *Giardia*, *Cryptosporidium* (from 1990 to 1998), total coliform and fecal coliforms, and turbidity (1998) are summarized in Table 3.1. Data for 1999 to 2002 are provided in Chapter 6.

Flow is measured at the United States Corps of Engineers 14 gage station located approximately 10 miles upstream of the water treatment plant at Davenport (Figure 3.3). Daily mean discharge data from 1993 through 1998 were retrieved from the United States Corps of Engineers (USCE) website. Descriptive statistics of these data are summarized in Table 3.1.

TREATMENT PLANT CHARACTERIZATION

The East River Station treatment plant serves 45,800 customers in the cities of Davenport, Bettendorf, Riverdale, Panorama Park, and Portions of Scott County along the Mississippi River in east-central Iowa. The source of supply for the East River Station is the Mississippi River. Raw water from the river is treated in the East River Station Plant by a variety of chemical and physical processes. Figure 3.4 shows the plant schematic for the Davenport water treatment plant. The treatment plant serves a population of about 135,000 people and is located at an elevation of 571 ft (174 m).

The 30 million gallon per day (MGD) East River plant was originally constructed in 1873, but was substantially upgraded in 1991 when the Superpulsator plant was built. The Mississippi River intake consists of two 3-foot square openings in the concrete seawall, one near the top and one near the bottom of the river profile, and a 30-inch pipeline that extends 430 feet from shore to the middle of the river channel. Each of the inlets may be operated independently. Water flows by gravity from the river through inlet chambers inside the sea wall through dual channels through twin traveling screens in the raw water pumping station. A set of three raw water pumps lifts water to the Superpulsators and conventional sedimentation basins. Pre-treatment disinfection with free chlorine is provided in two Superpulsator clarifiers and two conventional flocculation-sedimentation basins. The Superpulsator effluent flows by gravity to the filters. The sedimentation basin effluent is collected in a common fore bay and pumped by transfer pumps to the filters. There are 20 rapid-gravity filters. The utility uses granular activated carbon as a combination filtration/adsorption medium for removal of particles, organics, and for treatment of objectionable tastes and odors. Carbon media was first installed in East River Station in 1973 and is replaced or regenerated on an 18-month cycle. Filtered water flows by gravity to a below-ground clearwell. Ammonia is applied after filtration to form chloramines. During the study, a new 2.47 MG clearwell was completed in September 2001 to provide additional storage and disinfection contact time.

East River Station discharges its water treatment residuals to a Corp of Engineers interceptor storm sewer (the outfall of the storm sewer is located downstream of Lock & Dam 15, approximately one mile downstream of the treatment plant). Water treatment residuals are generated by the following waste streams: filter backwash water, sedimentation basin sludge pumps, Superpulsator blowdowns, and the periodic cleaning of the flocculator and sedimentation basins and the Superpulsator clarifiers.

There are two sets of distributive pumps, Main Service and High Service. Main Service encompasses the lower elevation area along the river and extending about a half mile back while High Service includes the higher elevation, larger, faster-growing part of the system. The distribution system includes more than 541 miles of mains and 5,053 hydrants and varies in elevation from 565 feet to 862 feet above mean sea level. In the Main Service pressure zone, there is one storage facility, the 3.18 MG Ripley Street Reservoir. There are five elevated storage tanks with combined capacity of 4.1 MG in the High Service zone. There are seven booster pumping stations in the distribution system that transfer water between areas or raise supplies to more distant higher elevation areas.

Iowa American is a voluntary member of the USEPA "Partnership for Safe Water" and received the Director's Award in 2001 for producing outstanding water quality.

Table 3.1
General watershed information for the Mississippi River at Davenport

Parameter		Summary Information
State		Iowa
Source type		River
Principal source water		Mississippi River
Watershed area (mi^2)		500 (HUC 07080101)
Vulnerability		recreational, agricultural, sewage, CSO, industrial
Point sources		Yes (>10 WWTP's upstream)
Non point sources		Yes (cattle, hogs, corn, soybeans)
Land use and land cover	% Urban	25
	% Agricultural	56
	% Rangeland	0
	% Forest	14
	% Others	5
Impervious surface (%)		25
	Served by the WTP	135000
Average Precipitation (in/year)		36.0
Elevation (ft/m)		571/174
Flow (cfs) [a] (min, mean, max)		15296, 67332, 267730
Giardia (cysts/100L) [b] (min, mean, max, # samples, % BDL)		0.0, 79.5, 909.1, 24, 88
Cryptosporidium (oocysts/100L) [b] (min, mean, max, # samples, % BDL)		0.0, 240.0, 2576, 24, 75
Fecal coliform (cfu/100mL) [b] (min, mean, max)		100, 577, 2600
Total coliform (cfu/100mL) [b] (min, mean, max)		20, 5977, 180000
Turbidity (ntu) [b] (min, mean, max)		1.0, 25.3, 588.0

x No data. BDL (below detection limit) results were set to 0.
[a] USGS data from 1993 to 1998. [b] Historical data, 1990-1998.

Figure 3.1. Mississippi River Watershed
(Source: Contaminants in the Mississippi River Report, 1987-1992, USGS)

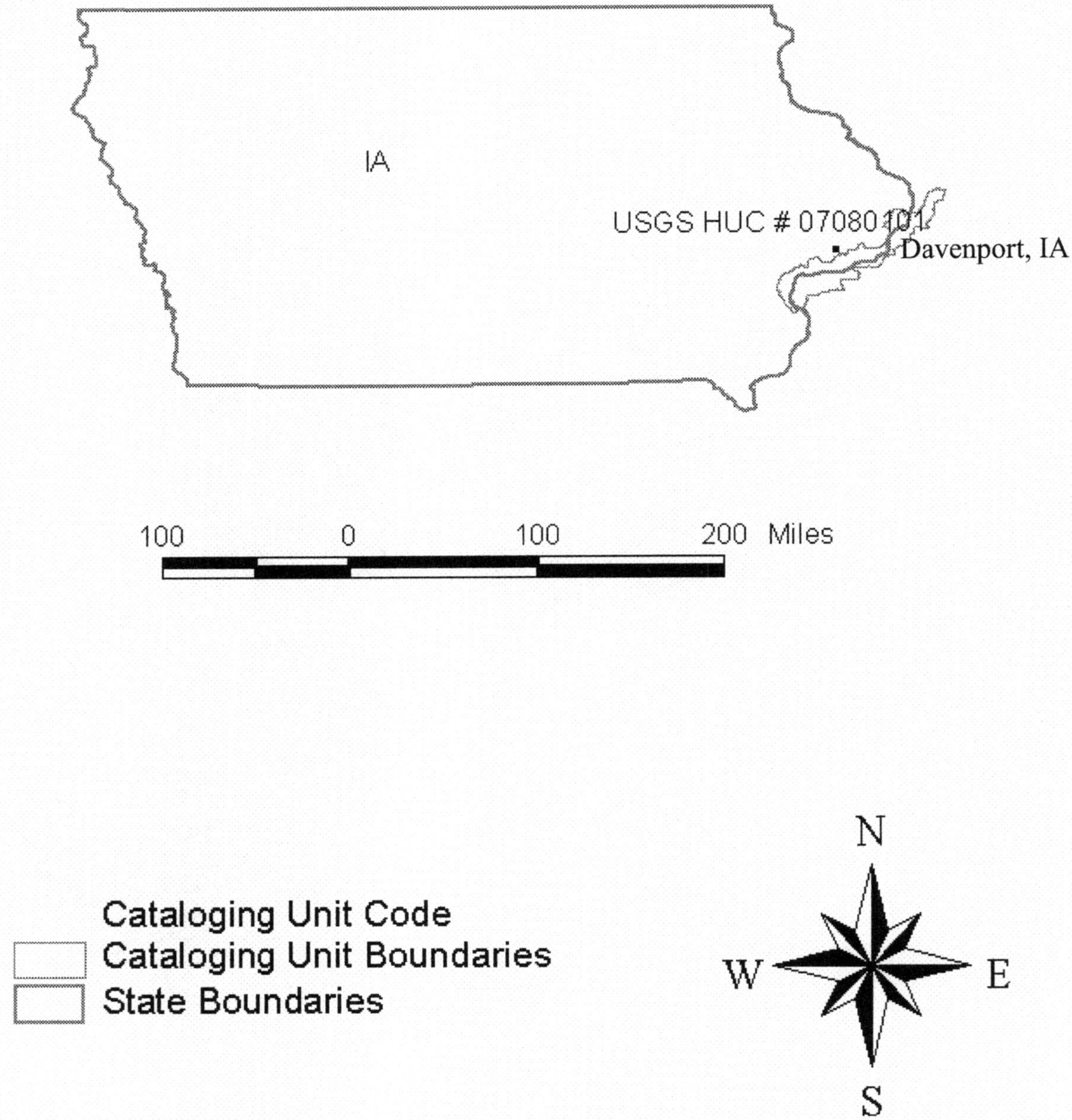

Figure 3.2. USGS HUC 07080101, Copperas-Duck
(Source: EPA BASINS Database, Version 2.0, 1998b)

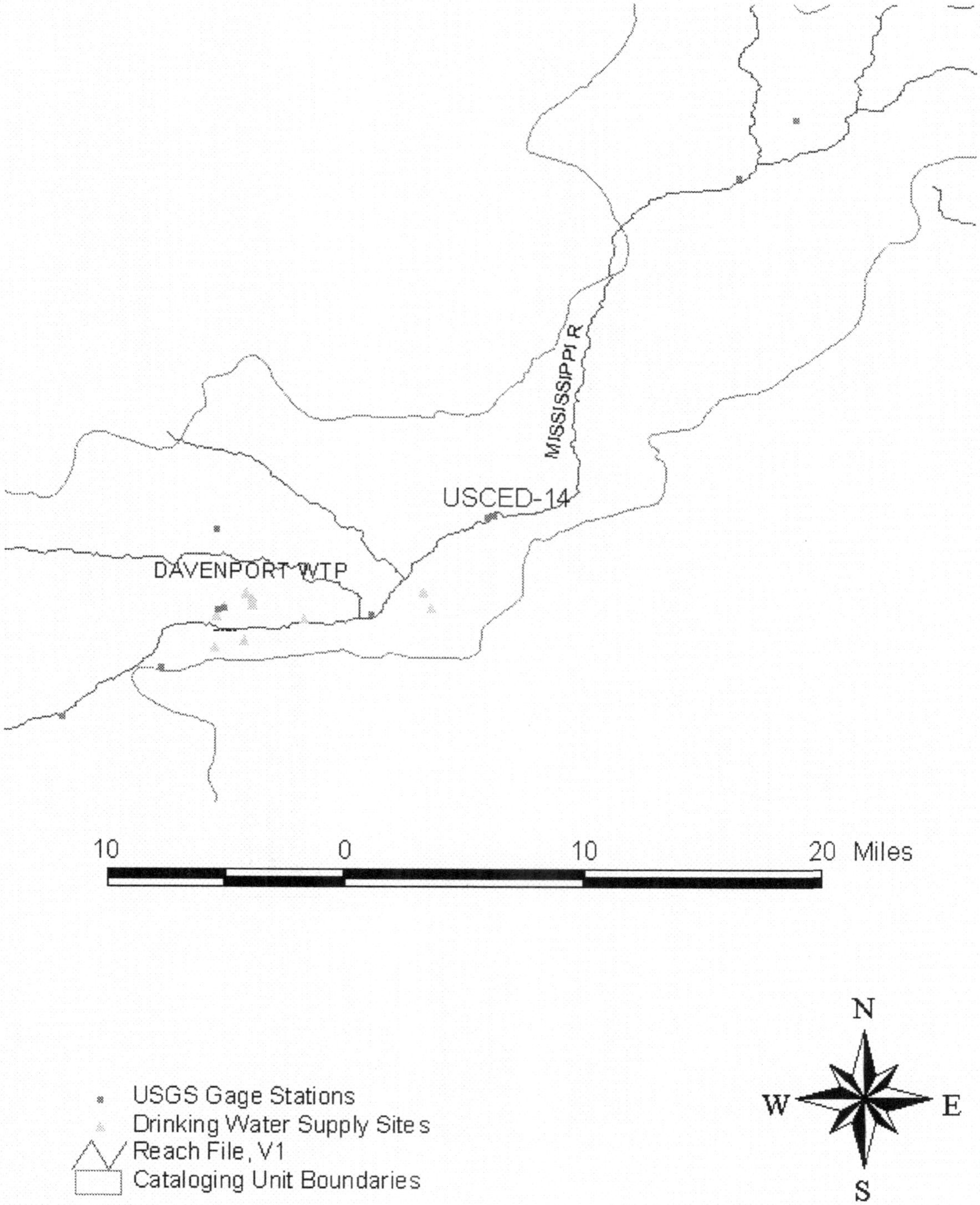

Figure 3.3. USCED-14 Gage Station
(Source: EPA BASINS Database, Version 2.0, 1998b)

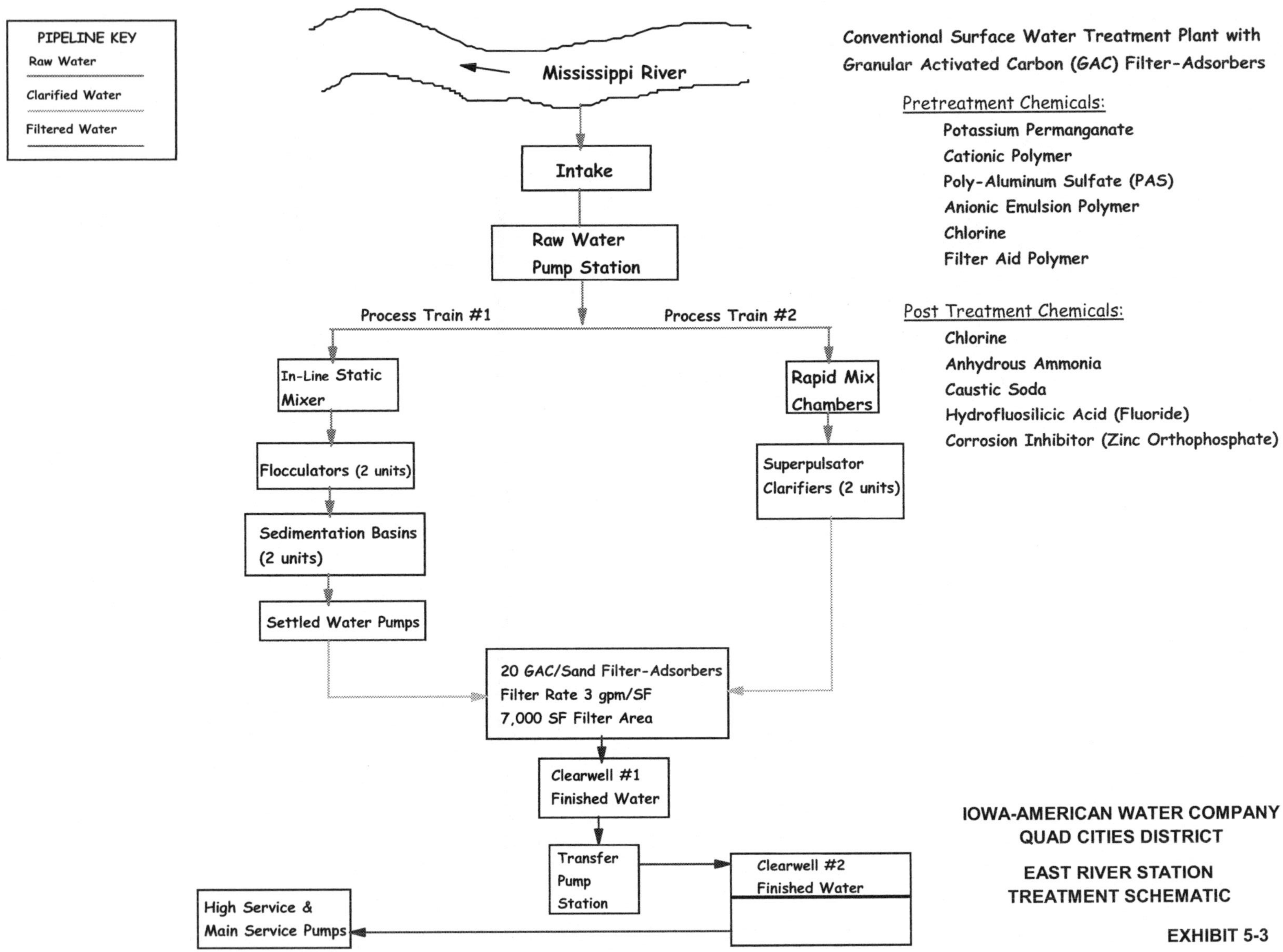

Figure 3.4. Schematic of the Iowa American Treatment Plant at Davenport

CHAPTER 4
MATERIALS AND METHODS

METHOD 1623

Description And Source Of *Cryptosporidium parvum* Oocysts

The Iowa isolate of *C. parvum* oocysts, which was originally obtained from Harley Moon (National Animal Disease Center, Ames, Iowa) and maintained at the Sterling Parasitology Laboratory (University of Arizona) was used throughout this study for evaluation of method performance from various source water matrices. This isolate has been well characterized and has been used in numerous studies including restriction fragment length polymorphism analysis where two separate passages through calves demonstrated identical banding patterns (Ortega et al. 1991). The Iowa isolate has been used to determine the *C. parvum* dose response in neonatal mice (Korich et al. 1990; Finch et al. 1993a) and has been used for measuring oocyst inactivation by various disinfectants, including chlorine, chlorine dioxide, monochloramine (Korich et al. 1990), ozone (Korich et al. 1990; Finch et al. 1993b, Bukhari et al. 2000), and ultraviolet light (Clancy 1998; Bukhari et al. 1999). Recently, this isolate has been used for determining infection and immunological characteristics in controlled experiments with healthy human volunteers (Dupont et al. 1995; Chappell et al. 1996; Okhuysen et al. 1998). Genetic analysis of the Iowa isolate from the Sterling Parasitology Laboratory over a 16-month period revealed that the monthly preparations were identical with a unique single nucleotide polymorphism (Aboytes and LeChevallier 2002). Highly purified suspensions of oocysts were obtained from the University of Arizona, where they are maintained by experimentally infecting neonatal calves, with protocols for calf care that were approved by the University of Arizona Institutional Animal Care and Use Committee. Stock oocysts were received in deionized water (DI) containing 0.01% Tween 20 and an antibiotic solution (100 units penicillin and 100 μg gentamicin/mL). Working suspensions were prepared and stored at 4°C in DI with no additional antibiotics added. The age of oocysts was less than three months old.

Description And Source Of *Giardia intestinalis* Cysts

Giardia intestinalis (the internationally accepted name for *G. lamblia*) cysts were obtained from Waterborne, Inc, New Orleans, and were originally isolated from an infected human. This strain is known as the H3 strain, which has been maintained by experimentally infecting Mongolian gerbils. Cysts were purified from gerbil feces by sucrose density gradient centrifugation followed by Percoll gradient purification. Purified cysts were stored in DI water and antibiotics (gentamicin, 0.1mg/mL; streptomicin, 0.1mg/mL; and penicillin, 100 U/mL) at 4°C. The age of cysts used in this study was less than one month.

Sample Collection And Elution Procedures

The utility collected source water samples weekly to be analyzed for *Cryptosporidium* oocysts and *Giardia* cysts, and indicator organisms. All source water samples were collected according to USEPA Method 1623 (USEPA 1999), using standard Envirochek filters (Pall-Gelman) at a flow rate of 2 liters per min, however, a minimum sample volume of 20 L was collected (10 L for direct IFA analyses and 10 L for CC-PCR analyses). Samples were transported on ice packs to Clancy Environmental Consultants, Inc (CEC) for elution. The elution procedure was as described in detail in Method 1623 (USEPA 1999). Briefly, a 120 mL volume of Laureth-12 buffer was added to the capsule, and each filter was placed on a mechanical shaker (Wrist Action Shaker, Labline Model 3589, Fisher Scientific) and shaken with the inlet valve in the 12 o'clock position for 5 min. The eluant was transferred to a 250 mL conical centrifuge tube and a further 120 mL of elution buffer was added to the capsule, which was shaken again with the inlet position at 4 o'clock for 5 min. Without decanting the eluant, the filter was rotated to the 8 o'clock position and shaken for an additional 5 minutes. This eluant was pooled with the eluant generated from the first wash, the entire sample (approximately 250 mL) was centrifuged (1500 x g; 15 min), if the packed pellet volume was 1 mL or less, the supernatant was aspirated to 10 mL and subjected to IMS. In the event the packed pellet volume exceeded 1 mL, the supernatant was aspirated to a volume where 10 mL of the resuspended volume contained 1 mL packed pellet volume. A subsample containing 1mL packed pellet volume was subjected to IMS.

Immunomagnetic Separation Procedure

The Dynal IMS (Dynabeads G/C combo IMS kits, Dynal A.S. P.O. Box 158 SkØyen, N-0212, Oslo, Norway) procedure was followed according to the manufacturer's instructions with a few exceptions. Briefly, each 10 mL volume of the test sample was placed in a screw cap Leighton tube, 1 mL 10x SL-buffer A, 1 mL 10x SL-buffer B and 100 μL of the anti-*Cryptosporidium* and 100 μL of the anti-*Giardia* bead conjugates were added. Each sample was be allowed to rotate through 360° for 1 hr (at room temperature) and the Leighton tube was placed in a magnetic particle concentrator (MPC)-1 to separate the bead-oocyst complex from the contaminating debris. The beads were resuspended in 1 mL of 1x SL-buffer A, transferred into an Eppendorf tube, separated by using a MPC-M and the supernatant removed and discarded. The beads were resuspended in 200 μL of Hanks balanced salt solution (HBSS) with bromophenol blue (0.001%) and 50% of the sample (100 μL) was shipped to American Water at the Belleville Lab for CC-PCR analysis. The remaining 100 μL of the sample, representing half of the original sample volume (i.e. 10 L) was subjected to dissociation using 0.1 N HCl, and in turn, the neutralization procedure was performed on the microscope slide, by using 10 μL of 1N NaOH respectively. Each sample concentrate was placed on an individual well of a three-welled microscope slide, dried (42°C), methanol fixed, labeled with an anti-*Cryptosporidium* and anti-*Giardia* fluoroscein isothiocynanate labeled monoclonal antibody (FITC-mAb) (Waterborne Inc., New Orleans, LA), and examined using epifluorescence microscopy.

Epifluorescence Microscopy

A Zeiss Axioskop fluorescence microscope (Carl Zeiss MicroImaging Inc, Thornwood, NY), equipped with blue filter block (excitation 490 nm; emission 510 nm) was used for detection of FITC-mAb labeled oocysts/cysts at x 360 magnification. Confirmation of oocysts/cysts was achieved at x 640 magnification by using a UV filter block (excitation 400 nm; emission 420 nm) for visualization of DAPI and internal morphology of oocysts/cysts was observed by using Nomarski- DIC microscopy.

Spiking Procedure For Quality Assurance

Additional quality control matrix samples for the purpose of spike recovery analysis were collected quarterly by the utility at the same time when the regularly scheduled samples were collected. Samples were collected using the same procedures as described above. These replicate samples were sent to Clancy Environmental Consultants where a known number of *C. parvum* oocysts (approximately 500) and *G. intestinalis* cysts (approximately 500) were spiked directly through the inlet port. These spiked samples were eluted from the capsule filters by using the same procedure as that described in the previous section and the 10 mL concentrate generated from each sample was subjected to IMS.

CELL CULTURE-POLYMERASE CHAIN REACTION METHOD

In Vitro Cell Culture Of *C. parvum*

Human ileocecal adenocarcinoma (HCT8; ATCC CCL244) cells were cultivated as previously described (Woods et al. 1995). Cell culture Maintenance Medium consisted of RPMI 1640 with L-glutamine (Gibco BRL, Grand Island, NY), 5% fetal bovine serum (FBS), pH 7.2, 15 mM HEPES buffer (N-[2-hydroxyethyl] piperazine-N'-[2-ethanesulfonic acid]), 100,000 U l^{-1} penicillin G, 100 mg l^{-1} streptomycin, 700 µg l^{-1} amphotericin B, and 12.5 mg l^{-1} tetracycline. Growth Medium used for the *in vitro* development of *Cryptosporidium parvum* contained 10% FBS, 15 mM HEPES buffer, 50 mM glucose, 35 mg l^{-1} ascorbic acid, 1.0 mg l^{-1} folic acid, 4.0 mg l^{-1} 4-aminobenzoic acid, 2.0 mg l^{-1} calcium pantothenate, 100,000 U l^{-1} penicillin G, 100 mg l^{-1} streptomycin, 700 µg l^{-1} amphotericin B, and 12.5 mg l^{-1} tetracycline. Twenty-four hours prior to inoculation, 96 well cell culture microplates were seeded with 5 x 10^4 HCT8 cells per well. Plates were then incubated at 37°C in a 5% CO_2 humidified incubator to allow for the development of 100% confluent monolayers.

Dissociation From Immunomagentic Separation Beads

Splits of IMS-purified (up to the dissociation step) water samples received from CEC were prepared for cell culture by adding 98 µl of acidified HBSS (AHBSS) containing 2% trypsin and 2.0 µl of 1 M HCl. Samples were then incubated at 37°C for 1 h with 10 sec of vortexing every 15 min. Following separation of the samples in the MPC-M concentrator, the supernatants containing the dissociated oocysts were transferred to microfuge tubes. To ensure recovery of oocysts, a second wash of the IMS beads with 100 µl of AHBSS/1% trypsin was performed. Pooled dissociated samples were neutralized with 0.5 N NaOH (approx. 4.0 µl), and centrifuged at maximum speed in a microcentrifuge for 2 min with no brake and aspirated down to 20 µl for CCPCR. Just prior to inoculation of monolayers, 50 µl of Maintenance Medium were removed. Samples for CCPCR were resuspended in 80 µl of prewarmed Growth Medium immediately following IMS dissociation and used to inoculate a HCT8 cell monolayer. After inoculation, monolayers were incubated at 37°C in a 5% CO_2 humidified incubator for 72 h prior to DNA extraction. AHBSS/1% trypsin-treated viable *C. parvum* oocysts were used as CC-PCR positive controls (approx. 5 to 20 oocysts per cell monolayers, depending on infectivity of oocyst stock) and single cycle freeze/thaw killed oocysts were used as CC-PCR negative controls. After incubation the cell monolayers were washed five times with 200 µl of PBS to remove unexcysted oocysts.

DNA Extraction From Oocysts And HCT-8 Cell Monolayers

Oocyst DNA was prepared using a rapid and simple freeze/thaw extraction procedure. Chelex 100 resin (Bio-Rad, Hercules, CA) was prepared as an equilibrated 1:1 volume-to-volume mix of resin and 1X Tris-EDTA buffer (1X TE, pH 8.0). For each sample an equal volume of prepared Chelex resin was added. Oocysts were lysed by eight cycles, 1 min each, with freezing in liquid nitrogen and thawing at 98 °C. DNA was purified from HCT-8 cells using the QIAamp DNA mini kit (Qiagen, Valencia, CA) and a modified extraction protocol. Modifications included incubation of the cell monolayers with lysis buffer for 15 min at 70-75 °C, and elution of DNA from the QIAamp DNA columns using 50 µl of 0.1X TE pH 8.0 at 70-75 °C for 5 min.

Cell Culture PCR (CC-PCR)

PCR primers specific for the *C. parvum hsp70* gene were used resulting in a 346 bp product. The primer sequences were as follows: forward primer CPHSPT2F 5′-TCCTCTGCCGTACAGGATCTCTTA-3′ and reverse primer CPHSPT2R 5′-TGCTGCTCTTACCAGTACTCTTATCA-3′. PCR was performed using a Perkin-Elmer model 9600 thermal cycler (PE Applied Biosystems, Foster City, CA). Each 100 µL PCR mixture contained 10.0 µL of 10x amplification buffer with Mg (1.5 mM final; Roche Molecular, Branchburg, NJ); 200 µM of each dATP, dTTP, dCTP, and dGTP (Amersham Pharmacia Biotech, Piscataway, NJ); 200 nM of each of forward and reverse CPHSP2 primer; 2.5 µL of 30 mg ml^{-1} bovine serum albumin (BSA; Sigma Chemical Co., St. Louis, MO); 5.0 U AmpliTaq Gold DNA polymerase (Roche Molecular) and varying amounts of *C. parvum* template DNA (up to 50 µL). Aliquots (up to 10% of the final PCR volume) of Chelex 100 freeze/thaw prepared oocyst DNA equivalent to approximately 10 oocysts were used as PCR positive controls and molecular grade water was used to prepare template-negative PCR controls. Amplification conditions were as follows: initial denaturation at 95°C for 10 min; 45 cycles of denaturation at 95°C for 30 sec and annealing at 60°C for 1 min; followed by a final extension at 72°C for 10 min and a 4°C hold. Amplification products were separated by horizontal gel electrophoresis on a 2.0 % agarose gel (Amresco, Solon, OH) containing 0.5 µg ml^{-1} ethidium bromide (Sigma Chemical Co., St. Louis, MO) and visualized under UV light. Gel images were captured using a gel documentation system (UVP, Inc., Upland, CA).

Quantitative Sequence Detection

Quantitative cell culture PCR using cell culture quantitative sequence detection (CC-QSD) was performed using the PE Applied Biosystems 7700 Sequence Detector (Perkin-Elmer, Foster City, CA). The primers were the same as described above and the TaqMan probe sequence was as follows: CPHSP2P2 5′-TGTTGCTCCATTATCACTCGGTTTAGA-3′. Each 100 µl CC-QSD mixture contained 10.0 µl of 10X TaqMan A amplification buffer (Roche Molecular, Branchburg, NJ), 1.5 mM $MgCl_2$, 200 µM of each dATP, dTTP, dCTP, and dGTP (Amersham Pharmacia Biotech, Piscataway, NJ); 200 nM of each forward and reverse CPHSPT2 primer and CPHSP2P2 TaqMan probe (Synthegen, Houston, TX); 5.0 μl of 30 mg ml^{-1} bovine serum albumin (BSA; Sigma Chemical Co., St. Louis, MO); and 5.0 U AmpliTaq Gold DNA polymerase (Roche Molecular). Purified cell culture DNA for each sample (50 μl), or aliquots of Chelex-treated oocyst DNA controls (up to 10 μl), was used as template. Human β-actin gene DNA and appropriate primers and probe (Roche Molecular) were used as amplification standards. Molecular grade water was used to prepare no template negative controls. Amplification conditions were as follows: initial denaturation at 95°C for 10 min; 45 cycles of denaturation at 95°C for 30 sec and annealing at 60°C for 1 min; followed by a final extension at 72°C for 10 min and a 4°C hold.

Genotyping of *C. parvum* Isolates

CÇPCR products were cloned and sequenced for confirmation of homology to the *C. parvum hsp70* gene. Products were cloned using a TOPO TA cloning kit (Invitrogen, Carlsbad, CA) according to the manufacturer's instructions. Cloned products were sequenced commercially (ACGT, Northbrook, IL and DNA Sequencing Service, The University of Arizona, Tucson, AZ) and sequence homology to the *C. parvum* KSU-1 *hsp70* gene (Khramtsov et al. 1995) was confirmed using Gene Runner version 3.0 (Hastings Software, Inc., Hastings, NY) or GeneBase Version 1 (Applied Maths, Kortrijk, Belgium). Sequences were aligned and similarity dendrograms using the unweighted pair group method and the arithmetic averages (UPGMA) method were generated using GeneBase. Duplicate clones of each PCR product were sequenced and analyzed to identify potential sequencing errors. When possible, clones obtained from independent CÇPCR of the same sample were analyzed to exclude random nucleotide misincorporation by *Taq* DNA polymerase.

CLOSTRIDIUM PERFRINGENS SPORE ANALYSIS

Clostridium perfringens is a strict obligate anaerobe found in fecal material. The standard test method (Armon and Payment, 1987) for the detection and enumeration of *C. perfringens* spores from water is by membrane filtration (MF). Two hundred milliliters of effluent or raw water were heat-treated at 60°C for 15 minutes. Volumes ranging from 25 ml to 200ml were passed through Microfunnel II 0.45-μm membrane filters (Pall-Gelman, Ann Arbor, MI) in duplicate. The smaller volumes were typically necessary for the raw water samples due to high organism numbers. The filters were placed on modified m-CP agar as described in the ICR Microbial Laboratory Manual (USEPA, 1996b), and incubated anaerobically for 24 hours at 44.5°C. After incubation the yellow straw colored colonies were exposed to ammonium hydroxide fumes. Colonies that turned pink or magenta colored were scored as *Clostridium perfringens*. The results were reported as colony forming units (CFU)/100 mL.

BACILLUS SUBTILIS SPORE ANALYSIS

Bacillus subtilis is an aerobic spore-forming bacterium, commonly found in water and soil. Two hundred milliliters of effluent or raw water was heat-treated at 60°C for 20 minutes. Duplicate 100 mL aliquots of effluent sample were passed through Microfunnel II 0.45 μm membrane filters (Pall-Gelman). The filters were placed on modified starch agar (MSA) and incubated for 20 hours at 37°C. Raw water samples were analyzed by spread plating volumes of

100 μL or less onto MSA plates. These were also incubated for 20 hours at 37°C. Colonies with a clear zone upon flooding with Lugol's iodine solution were scored as *Bacillus subtilis.* The results were reported as *Bacillus subtilis* CFU/100mL.

BACTERIOPHAGE ANALYSIS

Coliphage presence in water samples may be an indicator of fecal contamination. Analysis of effluent, raw and distribution water samples for male-specific (F^+) and somatic coliphage was performed by the single agar layer method as described in EPA Method 1602 (USEPA, 2001c). Briefly, 100 mL of water sample was mixed with $MgCl_2 \cdot 6H_2O$ (10 g), molten double strength Tryptic Soy Agar, and either *E. coli* host strain ATCC#700609 or ATCC#700891 before being poured into Petri dishes. The samples were incubated for 16-20 hours at 37°C. Samples containing coliphage were determined by observing circular lysis zones (plaques) within the bacterial lawn. Verification of plaques was completed by transferring any questionable plaques using a sterile cotton-tipped stick to Tryptic Soy Broth and analyzing a portion of this suspension by the double agar layer method according to EPA Method 1602 (USEPA 2001c). Results were reported as plaque forming units (PFU)/100 mL.

ENTERIC VIRUS ASSAY

Filter Processing

Monthly samples of source water and plant effluent were collected and processed for culturable enteric viruses using the procedures described in the ICR Microbial Laboratory Manual (USEPA, 1996b). Water samples were obtained by passing a minimum of 400 gallons of effluent water, or 60 gallons of raw water, through a 1 MDS filter (CUNO Inc., Meriden, CT), at a flow rate of 4 gallons per minute. Samples were shipped on ice via overnight delivery to the American Water central laboratory in Belleville, Illinois. The filters were eluted using 1L of an autoclaved solution of 1.5% beef extract, 0.05M glycine, pH 9.5 (BEG). The BEG was left to contact the filter for 1 minute and then expelled using nitrogen gas (N_2) into a sterile beaker. The eluate was then poured back into the filter housing and again forced out using N_2 as before into the same beaker. After the second elution, the pH of the BEG was adjusted to 3.5 using 1M HCl and stirred for 30 minutes. The pH was monitored continuously and not allowed to fall below pH 3.4. After 30 minutes the solution was centrifuged at 3,290 x g at 4°C for 15 minutes with no brake. The supernatant was discarded and the resulting pellets resuspended in a final volume of 30mls of 0.15M Na_2HPO_4 (pH 9.4). This solution was adjusted to pH 9.0-9.5 using 1M NaOH. At this point the solution was centrifuged again at 3,290 x g at 4°C for 15 minutes with no brake. The supernatant was saved and adjusted to pH 7.0-7.5 using 1M HCl. The adjusted supernatant was filtered though a 0.22 micron filter and the final volume recorded before storing at -80°C for use in the cell culture analysis.

Cell Culture Analysis

Cell culture analysis was performed according to the ICR protocol (USEPA, 1996b). Buffalo Green Monkey (BGM) kidney cells were grown to confluent monolayers in 25 cm^2 plastic flasks using BGM growth medium. BGM growth medium consisted of Minimal Essential

Media with Earle salts (Invitrogen, Rockville, MD) containing 10 % fetal bovine serum (Hyclone, Logan, UT). 1-mL aliquots of the previously stored sample concentrates were added to twenty 25cm^2 flasks. The BGM flasks were incubated for 60 minutes at 37°C with rocking every 15 minutes to aid in virus adsorption. Fresh maintenance media consisting of Minimal Essential Media with Earle salts and 2 % fetal bovine serum was added to the flasks. The flasks were incubated at 37°C and monitored for viral cytopathic effect (CPE) for 14 days. All flasks were confirmed positive or negative for CPE by inoculation of the media onto a fresh monolayer of BGM cells and observed an additional 14 days. Samples that produced positive or negative CPE after two passages were confirmed by assaying the cell culture harvests by reverse transcription polymerase chain reaction (RT-PCR).

Virus Analysis by Reverse Transcription Polymerase Chain Reaction

Cell cultures (with medium) from flasks exhibiting viral cytopathic effect (CPE), along with a positive and four negative control flasks, were lysed by three -80°C/ 37°C freeze/thaw cycles. Nucleic acids from the lysates were extracted using the QIAamp Viral RNA Kit (Qiagen, Valencia, CA) according to the manufacturer's directions. The nucleic acids were examined for the presence of enteroviruses and reoviruses by RT-PCR. For negative cell culture samples, after three freeze/ thaw cycle, 100 µl of cell culture harvest from each of the 10 flasks were pooled together. The pooled cell culture harvest was used for nucleic acid extraction by QIAamp kit and subsequent RT-PCR as described for positive cell culture samples.

The primer set used for the detection of reoviruses was REOL3 (REOL3F: 5' -CAG TCG ACA CAT TTG TGG TC – 3' and REOL3R: 5'-GCG TAC TGA CGT GGA TCA TA-3'; Spinner and DiGiovanni, 2001). The primer ser used for the detection of enteroviruses was PAN (PAN1: 5'-CCT CCG GCC CCT GAA TG -3' and PAN2: 5'-ACC GGA TGG CCA ATC CAA-3'; Abbaszadegan el al, 1999).

Each one-tube 50 µL RT-PCR cocktail contained: 10 µL of purified nucleic acid sample, 10 mM Tris, pH 8.3, 50 mM KCl, 2 mM $MgCl_2$, 200 µM each of dATP, dTTP, dCTP, and dGTP (Pharmacia BioTech, Arlington Heights, IL), 300 nM of each forward and reverse primer for either PAN or REOL3 primers (Gibco BRL, Grand Island, NY), 2.5 U AmpliTaq Gold DNA polymerase (Applied Biosystems, Foster City, CA), 5 U MuLV reverse transcriptase (Applied Biosystems), and 20 U RNase inhibitor (Applied Biosystems). RT-PCR cocktails were overlaid with mineral oil and run on a Stratagene (La Jolla, CA) Robocycler 96 thermal cycler. Reverse transcription was carried out at 42°C for 30 min; followed by an initial denaturation at 95°C for 10 min., heat inactivation of MuLV RT and activation of AmpliTaq Gold. Amplification consisted of 40 cycles of denaturation at 95°C for 30 sec, annealing at 55°C for 1 min, and extension at 72°C for 30 sec; followed by a single final extension at 72°C for 7 min and a 4°C hold.

Amplification products were separated by horizontal gel electrophoresis on 2.0% agarose gels (Amresco, Solon, OH) containing 0.5 µg/ml ethidium bromide. The gels were electrophoresed for 1.5 h at a constant voltage of 100 V and were visualized under UV light. Gel images were captured with a gel documentation system (UVP, Inc., Upland CA).

Sequence Analysis

The RT-PCR products were cloned using TOPO TA cloning kit (Invitrogen, Carlsbad, CA) and the purified products were sent to the University of Arizona for sequencing. The sequence data were compared to sequences in GenBank (an international public sequence database) using the web-based method BLAST (Basic Local Alignment Search Tool) search. Four of the eight positive water samples were confirmed by sequencing.

ADDITIONAL WATER QUALITY AND QUANTITY DATA

Additional water quality parameters were collected by the participating utility using State of Iowa approved procedures (American Public Health Association 1998). A summary of the sampling schedules and test parameters is shown in Appendix B. Data for on-line instruments are shown in Table 4.1. Average daily values were calculated for on-line information and hourly operational data. For example, although filter effluent turbidities were measured continuously, plant operators recorded hourly values and these hourly values were averaged to produce a daily average. Total and fecal coliforms were enumerated by the membrane filtration method using m-Endo, and m-FC agar, respectively (American Public Health Association 1998). Total coliform positive samples were tested for the presence of *E. coli* using the EC-MUG method (American Public Health Association 1998). Ancillary information was received from operational logs maintained by the Iowa-American Water Company.

DATA MANAGEMENT

Physical, chemical, biological and plant operational data for the Mississippi River and East River Station at Davenport, IA were collected and processed by three teams including the water utility, American Water (AW) and Clancy Environmental Consultants (CEC). The organization and management of the collected data was performed by Technology Planning and Management Corporation (TPMC) and based on Excel spreadsheets for data input, e-mail for data transmission and a relational database for data compilation. Construction and implementation of the database in Microsoft (MS) Access format allowed for data compilation, quality assurance, information management and data dissemination. Data dissemination to the participants in the study used a password protected website which was updated monthly.

Data Collection, QA/QC And Storage

TPMC received once a month pathogen and microbial indicator data from the two laboratories and monitoring and operational data from the Davenport utility. Data were assessed and evaluated for outliers, accuracy and completeness before compilation and storage.

Sample Identification

Environmental samples collected from the raw water, several locations within the treatment plant, distribution system and households were identified by a standard eleven-digit code. Figure 4.1 shows a graphical representation of the sample identification code. The first five digits of this code specified the collection date represented by the Julian calendar (Table A.1,

Table 4.1
Summary of on-line instrumentation for the East River Station, Davenport, Iowa

Process Site	WQ Parameter	Equip. Mfg.	Equip. Model No.	Units of Measure	Instrument Set-up Range of Measurement	SCADA No. Decimals Readout	High Alarm Set Point	High High Alarm Set Point	Low Alarm Set Point	Low Low Alarm Set Point
Raw Water	Turbidity	Hach	Surface Scatter 6	NTU	0 to 100 or 0 to 1000	0	50	100	2	1
	pH	Leeds & Northrup	7082 / 7773 probe	pH	0 to 14	0.00	9.00	9.50	7.30	7.00
	Temperature	Bristol Babcock	Series 97421/ 2508	F	0 to 100	0.0	87	90	32.5	32
Superpulsator Effluent	Turbidity	Hach	Surface Scatter 6	NTU	0 to 10	0.00	3.50	5.00	0.50	0.30
	Turbidity	Hach	1720C	NTU	0 to 10	0.00	3.50	5.00	0.50	0.30
	pH	Leeds & Northrup	7082 / 7773 probe	pH	0 to 14	0.00	8.00	8.50	6.90	6.80
	Free Cl2 Residual	W&T	Micro 2000	mg/l	0 to 5	0.00	4.00	3.80	1.00	0.50
	Particle Count	Hyac Royco(IBR)	VCOL25	um	2 to 25	0.000				
Sed. Basin #1 Effluent	Turbidity	Hach	1720C	NTU	0 to 10	0.00	3.50	5.00	0.50	0.30
	pH	Leeds & Northrup	7082 / 7773 probe	pH	0 to 14	0.00	8.00	8.50	6.90	6.80
	Free Cl2 Residual	W&T	Micro 2000	mg/l	0 to 5	0.00	4.00	3.80	1.00	0.50
Sed. Basin #2 Effluent	Turbidity	Hach	1720C	NTU	0 to 10	0.00	3.50	5.00	0.50	0.30
	pH	Leeds & Northrup	7082 / 7773 probe	pH	0 to 14	0.00	8.00	8.50	6.90	6.80
	Free Cl2 Residual	W&T	Micro 2000	mg/l	0 to 5	0.00	4.00	3.80	1.00	0.50
Individual Filter Effluent(s)	Turbidity	Hach	1720C	NTU	0 to 1	0.00	0.20	0.30	0.03	0.01
	Loss of Head			Ft.	0 to 10	0.0	6.0	7.0	0.2	0.1
	Rate of Flow	Primary Flow Signal	HVT-CI	MGD	0 to 2.00	0.00	1.80	2.00	0.20	0.10
	Filter Run Hours			Hrs.	0 to 100	0.0	48	53		
Clearwell Effluent (Prior to Ammonia feed pt.)	Free Cl2 Residual	W&T	Micro 2000	mg/l	0 to 5	0.00	3.00	3.50	1.75	1.50
Finished Water (Plant Effluent)	Turbidity	Hach	1720C	NTU	0 to 1	0.000	0.090	0.150	0.030	0.010
	pH	Leeds & Northrup	7082-11 / 7773	pH	0 to 14	0.00	7.40	7.50	7.25	7.20
	Total Cl2 Residual	W&T	Micro 2000	mg/l	0 to 5	0.00	3.00	3.50	1.75	1.50
	Total Cl2 Residual	Hach	CL17	mg/l	0 to 5	0.00	3.00	3.50	1.75	1.50
	Fluoride	Leeds & Northrup	7082 / 7773 probe	mg/l	0 to 5	0.00	1.30	1.50	0.90	0.80
	Phosphate (Ortho)	Hach	Series 5000 Low Range	mg/l as PO4	0 to 5	0.00	1.40	1.50	1.10	1.00
	Temperature	Bristol Babcock	Series 97421/ 2508	F	0 to 100	0.0	87	90	32.5	32
	Particle Count	Hyac Royco(IBR)	VCOL60	um	2 to 25	0.000				

Appendix A). The first three digits of the Julian calendar represented the day and the other two digits represented the year.

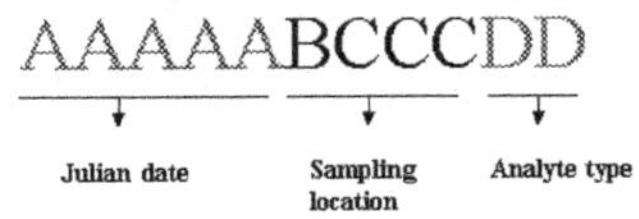

Figure 4.1. Sample identification code

The next four digits of the sample code identified the sample location. As shown in the first column of Table 4.2, *Sampling Location Code*, the first of these four digits was an alphanumeric character that identified the general location. The subsequent three digits were numeric and identified the specific site within a general location (Table 4.2, *Sampling sub-location code* column).

Table 4.2
Sampling location code and type

Sampling Location Code	Sampling Location Type	Sampling Sub-Location Code	Sampling Sub-Location Type
R	Raw water	000	None
P	Any location within the plant except effluent	000	None
E	Plant effluent	000	None
A, B, C, D	Four quarters of the distribution system	001 to 032	Different locations in the distribution system
T	Household tap	001 to 400	Number of the household to be surveyed
V	Household treatment device	001 to 400	Number of the household to be surveyed

The last two digits of the sample identification code specified the analyte type. Table 4.3 lists analyte code and correspondent type and location from where the analyte was collected.

Table 4.3
Analyte code and type

Analyte Code	Analyte Type	Sampling Location
GC	*Cryptosporidium*, *Giardia* and algae to be measured by USEPA Method 1623	Treatment plant influent
CL	*Clostridium perfringens*	Treatment plant influent and effluent
CC	CC-PCR	
BS	*Bacillus subtilis*	
CV	Culturable viruses	
PG	Mean male specific and mean somatic coliphage	Treatment plant influent, effluent and distribution system
WQ	Particle count, turbidity, alkalinity, pH, hardness, fecal and total coliform, temperature, fluoride, phosphate, free and total chlorine, HPC bacteria, effluent water flow, CT obtained/CT required, river state, river flow, precipitation	Different locations within the treatment plant
WQ	Total coliform, total/free chlorine, pH, turbidity, phosphate, temperature, HPC bacteria	Distribution system
WQ	Total coliform, HPC bacteria, total chlorine, pH, turbidity, temperature	Household tap and treatment device
PO	Plant flow rate, headloss, number of filters washed, average filter run, wash water flow rate, chemical dosages	Filters and other locations within the treatment plant

As an example to illustrate the sample identification used in this study, different samples were collected on November 1, 2000 at the influent and shipped to CEC for *Giardia* and *Cryptosporidium* analyses and American Water for *Bacillus subtilis*. These samples were labeled by the utility personnel as follows. The *Giardia* and *Cryptosporidium* sample identification code was 30600R000GC while the sample ID for *Bacillus subtilis* was 30600R000BS. In this manner, the sample labeling was clear enough to indicate when and where the sample was collected and the analytical test that needed to be performed. Two analyte codes, WQ and PO, were used by the East River Station personnel at the time of submitting data on a month basis to TPMC. In these two cases, a single code was used to describe a variety of environmental parameters (Table 4.3). Following the same example used for the biological sample identification, the sample 30600R000WQ provided raw water quality data including temperature, pH, alkalinity, hardness, total and fecal coliform, turbidity and other physical parameters.

Sample and Data Tracking

The personnel at the East River Station collected and labeled the sample with the unique sample code that has been designed specifically for this project. The sample arrived at CEC and American Water with a transfer form from the East River Station indicating the sampling date, sampling personnel, and qualifiers with respect to the sampling event (Figures B.1 and B.2, Appendix B).

In order to assess the two *Cryptosporidium* detection methods (Method 1623 and cell culture/PCR) samples received at CEC were split, with one portion assayed by Method 1623 at CEC and the other portion sent to American Water for Cell Culture-Polymerase Chain Reaction analysis (CC-PCR). The sample transfer form shown in Figure B.3 (Appendix B) was used to track the sample from CEC to American Water. Sample codes as discussed previously were unique and did not change with shipping except the analyte type. For example, when CEC sent the sample labeled 30600R000GC to American Water for CC-PCR analyses, the sample was recorded and labeled as 30600R000CC. Codes used by American Water and CEC are listed in Table 4.3.

Each laboratory maintained a bench sheet that included sufficient information for sample tracking and identification. Examples of laboratory bench sheets from CEC are shown in Appendix C, Figures C.1 and C.2. Information maintained in the bench sheets included sampling and receiving date, condition of sample, and sample ID. This facilitated sample tracking in the case of anomalous values for the parameter in question.

Data Collection

The prearranged data sheets for biological analyses results (Figures D.1 to D.6, Appendix D) and water quality measurements (Figures D.7 to D. 12, Appendix D) were supplied in advance to the laboratories and water utility. Data collection sheets including the sample ID's were designed and provided by TPMC to the East River Station. Data collection sheets used by the water utility were locked so that the pre-designed sample ID's and sampling date field could not be modified.

All of the data listed and detailed in Appendix D were sent to TPMC in MS Excel format for database entry. These data were checked for completeness, accuracy, and technical reasonableness by staff at TPMC. Data review and QA/QC procedures by TPMC included visual inspection of all data for obvious outliers and a determination if these values warrant corroboration from the originating source. Any discrepancy found in the data was discussed in detail with the utility and laboratories and corrected if necessary.

Data Storage

The data received in the Excel spreadsheets from the laboratories and water utility were then loaded into a Microsoft Access database designed specifically for this project using an automated procedure that employed temporary tables and transformation queries. Figure E.1 shows the Entity Relationship Diagram (ERD) for this database. Figures E.2 through E.13 show the specifics for each of the main entities (tables) shown in the ERD. Within each table or entity in Figures E.2 through E.13, a data dictionary is provided to aid users in understanding the structure of the database. Attributes are specified by name, unit, format, minimum, maximum, and description. Attribute is the field name for a variable that is listed in the database. Character fields are identified by the letter 'C'. For example, an entry that is noted as C (10) indicates a field with a maximum width of 10 characters. Numeric fields are denoted by an 'N' and date fields are labeled as 'dd/mm/yyyy'. A field, which is defined as numeric, will contain only numbers and is specified as N. An entry such as N (5.2) specifies a numeric entry with 2 decimal places and a maximum of 5 numbers (e.g. 100.00). The units for all numeric fields are specified in the Units column. In addition, proposed minimum and maximum values for the fields were indicated. All fields have been given preliminary entries (minimum and maximum) that describe the information content allowable for that entry within the database. A short description for each attribute was provided in the final column of each table.

Any data received by TPMC that did not fit the field designation or width were excluded from entry into the database and flagged by the data entry program. These data were inspected visually and appropriate action was taken to remedy the problem. This action likely involved contacting the source of the out of bounds data and constituted another step in the QA/QC procedure. Allowance was made for unanticipated outliers and some of the fields had to be modified in an on-going process as data were collected. This was anticipated and modifying field widths was generally a simple procedure.

Preliminary Data Analysis And Data Dissemination

Once the data were loaded into the database and quality assurance was completed, the contents of the database were transferred to SAS (Statistical Analysis System, SAS Institute Inc.) for further analyses. Preliminary analyses of the data received on a monthly basis were performed for each site. Descriptive statistics included sample size, mean, standard deviation, minimum, maximum, and 3 other percentiles (median, 90% and 95% percentiles). Physiochemical and biological data and descriptive statistics were posted on a restricted TPMC website (http://www.drinkingh2o.com/2580/index.htm). The site had a straightforward design with a total number of six pages where the data could be easily accessed and selectively downloaded. Besides the Home page, the website included the Database Page and another four pages. Through the Database Page the Principal Investigators (PI's) followed closely the sample collection since the page displayed a table where the cumulative number of samples collected and detection frequency were shown for all biological parameters monitored. The MS Access AwwaRF Project 2580 database could be also downloaded from the Database Page and was available to the PI's only. The remaining four other pages showed descriptive statistics for the most current month for each of the four monitored locations (raw water, effluent, distribution system, and households). From these four pages, the current and the previous month data with the descriptive statistics were also available for downloading. Each group including the project team, AwwaRF and water utility had its own password to access the site. The website was updated on a monthly basis as soon data from laboratories and water utility were received and processed at TPMC.

CHAPTER 5
QUALITY CONTROL

PROTOZOAN RECOVERY EFFICIENCIES

Enumeration Procedures Used In Spiking Studies

Enumeration Using A Hemacytometer

Stock suspensions of concentrated *Cryptosporidium* oocysts or *Giardia intestinalis* must be enumerated and diluted before the diluted suspension is used to spike samples. The oocyst /cyst concentration in the stock suspension was determined using a hemacytometer and involved application of a clean cover glass to the hemacytometer (Hauser Scientific 3200) followed by loading of each hemacytometer chamber with 8-10 µL of the vortexed oocyst/cyst suspension. Care was exercised in ensuring the entire chamber was filled without bubbles or overflowing of liquid into the surrounding moats. Once the chambers had been filled, care was taken to ensure cover glass was not adjusted or in any way disturbed. The filled hemacytometer was placed onto a microscope stage and the *Cryptosporidium* oocysts/*Giardia* cysts allowed to settle for 2 min before enumeration, using differential interference contrast microscopy (DIC) at a magnification of 400 - 640X.

Enumeration By Using Immunofluorescence

Immunofluorescence microscopy was used for determining the actual oocyst/cyst spike doses, for determining over-all method recoveries post-spiking and for detection of oocysts/cysts recovered from source water samples. For the diluted oocyst/cyst stocks, the vial containing the diluted suspensions was vortexed for 2 min and 100 µL aliquots from the suspension were applied to the center of a well of 12-mm-diameter three well slides (Meridian Diagnostics). Before removing subsequent aliquots, the vial was capped and gently inverted three times to ensure organisms were in suspension. Ten wells were prepared and counted, and the counts averaged to enumerate the spike dose. For environmental water sample concentrates, the volume post-IMS was adjusted to 200 µL, and 50% of the sample concentrate was applied onto an individual well of a 12-mm-diameter three well slide.

Individual slides containing either the diluted oocyst/cyst suspensions, environmental source water concentrates or spiked source water concentrates were dried (at 42°C) on a slide-warming tray, evaporated to dryness (approximately 1-2 hours) and 50 µL of absolute methanol was placed on each well containing the dried sample. The methanol was evaporated dry (approximately 3 - 5 min.) and then manufacturer's instructions were followed in preparing dilutions of anti-*Cryptosporidium* sp/anti-*Giardia* sp FITC-monoclonal antibody ([FITC-mAb] e.g. Crypto-a-Glo and *Giard*-a-Glo Waterborne, Inc.). An aliquot of the working dilution of the FITC-mAb (50 µL) was overlayed onto each of the 10 spike-dose wells, the slides placed in a humid chamber and incubated at 37 °C for 30 min.

Using a clean Pasteur pipette tip attached to a vacuum source, excess FITC-mAb was aspirated from the side of each well. A 50 µL aliquot of DAPI solution (1/5000 dilution in PBS of 2 mg/mL DAPI prepared in methanol) was applied onto each well, allowed to stand at room temperature for 2 min and the excess DAPI solution removed by aspiration. The FITC-mAb-labeled samples were stored in a dry box until dry and approximately 10 µL of mounting medium (60% glycerol, 40% PBS) containing an anti-quenching agent (e.g. 2% DABCO) was applied to the center of each well. A 22 x 50 mm coverslip was placed on each three-well microscope slide then the edges were sealed with clear nail polish. The slides were stored in a dry box, in the dark, until ready for microscopic analysis.

Spiked Sample Recoveries

To measure the efficiency of protozoan detection, 9 environmental samples were collected and separately spiked at the CEC laboratory and splits were sent to the American Water laboratory in Belleville, Illinois. Recovery efficiencies from both laboratories for these spiked samples are provided in Table 5.1. Two samples were not analyzed by CC-PCR due to shipping problems or failed PCR reactions. Spike dose suspensions used for one spiked sample were incorrectly prepared resulting in very low concentrations of oocysts and cysts being added to the capsule filter. Results from these spiking trials are considered invalid. The average percent recovery and standard deviation for *Giardia* and *Cryptosporidium* using Method 1623 was 38.0 ± 26.8% and 52.7 ± 35.0%, respectively. These results were within accepted performance levels for the method (USEPA 1999).

Because the CC-PCR method is a presence/absence technique, quantitative sequence detection (QSD) using the Perkin Elmer 7700 Quantitative Sequence Detector (QSD) to perform a real-time quantitative PCR was used to enumerate oocyst levels in the quality assurance samples. Based on quantitative PCR standard curves for known amounts of nucleic acids and oocyst stock CC-QSD baselines, the CC-PCR recovery efficiencies of spiked samples was determined. The average percent recovery and standard deviation for *Cryptosporidium* using the CC-QSD method was 52.2 ± 39.2%. There was no statistical difference ($p<0.05$) between Method 1623 and the CC-QSD results for *Cryptosporidium*.

Table 5.1
Recovery efficiency for *Giardia* and *Cryptosporidium* by Method 1623 and CC-QSD

QC Replicate	Method 1623 *Giardia* Recovery (%)	Method 1623 *Cryptosporidium* Recovery (%)	CC-QSD [2] *Cryptosporidium* Recovery (%)
1	12.9	15.8	55.3
2	26.7	35.7	NA [3]
3	11.9	8.0	43.7
4	1.2 [1]	0 [1]	19.3
5	79.8	54.5	130
6	32.9	63.2	59.2
7	59.7	102.8	8.2
8	14.0	42.2	NA [3]
9	65.7	99.4	50.0
Average	38.0	52.7	52.2

[1] Invalid due to low spiking dose.
[2] Quantitative Sequence Detection.
[3] QSD qualifiers: NA= sample not analyzed due to failed PCR reaction or out of holding time.

RECOVERY OF ENTERIC VIRUSES

The effect of filter elution procedures (beef extract-glycine, BEG), and buffers (sodium phosphate and phosphate buffered saline, PBS) used for virus filter processing was examined using spiked samples (one liter of BEG solution was spiked with $1.02x10^5$ PFU of poliovirus 3), concentrated using the organic flocculation procedure as described by the ICR protocol (USEPA 1996b), and enumerated using the plaque assay on BGM cells. The average recovery of viruses from BEG using sodium phosphate buffer and PBS were 53% and 50%, respectively (Table 5.2), which met acceptable recovery rates as described in the ICR (USEPA 1996b). The data suggest that there was no significant difference between the buffers with regard to recovery of viruses from BEG. Therefore, sodium phosphate buffer was used throughout the course of the experiment.

Matrix spike experiments were conducted on three separate occasions to determine the recovery efficiency of the enteric virus method for raw and plant effluent water samples. Samples were spiked with 200 plaque-forming units (PFU) of poliovirus LH type 3, in 40 L of water (for trial 1, 2,000 PFU were used). Samples were processed as outlined in the Materials and Methods section. Recovery efficiencies averaged 47.1% for raw water and 17.9% for plant effluent samples (Table 5.3). The low recovery of viruses in the effluent samples compared to raw water may have been due to the presence of background viruses in raw water, or the chemical characteristics of the effluent including soluble organic matter that may have reduced the virus adsorption efficiency (Sobsey and Glass, 1984). However, the results for the raw and plant effluent samples were within accepted performance levels for the method.

Table 5.2
Recovery efficiency of enteric viruses from spiked beef extract-glycine

Buffer	% Recovery		
	Trial 1	Trial 2	Average
Phosphate buffered saline	43	57	50
Sodium phosphate	48	58	53

Samples were spiked with $1.02x10^5$ PFU of Poliovirus 3

Table 5.3
Recovery efficiency of enteric viruses from raw and effluent water samples

Water Type	% Recovery			
	Trial 1	Trial 2	Trial 3	Average
Raw	UD*	31.4	62.8	47.1
Plant effluent	26.3	15.7	11.8	17.9
Negative control	0	0	No Data	--

*UD = Undetermined; every cell culture flask was positive for CPE; trial 1 spiking concentration was 10 times higher than subsequent trials.

RECOVERY OF INDICATOR ORGANISMS

Spiked recovery efficiency results for *Bacillus subtilis* and *Clostridium perfringens* spores are shown in Tables 5.4 and 5.5, and phage (the somatic phage, Phi-x-174; and male-specific phage, MS2) recoveries are shown in Tables 5.6 and 5.7. *Bacillus* spore recoveries average 95% while *Clostridium* spore recovery efficiencies averaged 92%. Although USEPA guidelines call for determination of recovery efficiencies of phage using the double agar overlay (DAL) method (USEPA, 2001b), recoveries were better when the single agar overlay (SAL) method was used (Tables 5.6 and 5.7). Considering that the SAL method was used for water analysis in this study, it seemed appropriate to report these results (120% for detection of Phi-x-174, and 106% for MS2). There appeared to be no effect on recovery efficiency due to the sample matrix (raw, plant effluent, distribution system, or laboratory-grade deionized water).

Table 5.4
Recovery efficiency for *Bacillus subtilis* spores

Date	Sample Matrix	Percent Recovery
09/25/2000	Effluent	72
09/25/2000	Effluent	89
09/25/2000	Effluent	101
11/02/2000	Effluent	95
11/02/2000	Raw	103
11/09/2000	Effluent	110
Average		95

Table 5.5
Recovery efficiency for *Clostridium perfringens* spores

Date	Sample Matrix	Percent Recovery
09/29/2000	Effluent	87
09/29/2000	Effluent	79
09/29/2000	Effluent	110
09/29/2000	Effluent	90
09/29/2000	DI water	93
Average		92

Table 5.6
Recovery Efficiency for Phi-x-174 Phage

Sample	Actual PFU spiked (via DAL)	Sterile DI Water	Matrix spike	
		% recovery based on DAL assay	% recovery based on DAL assay	% recovery based on SAL assay
DI water	29.5	188	NA	NA
DI water	122.5	124	NA	NA
Distribution	73.5	171	92	54
Distribution	87.5	132	119	90
Distribution	94	85	122	144
Distribution	131.5	109	135	124
Distribution	130	92	97	106
Distribution	86	87	90	103
Distribution	93.5	96	140	146
Distribution	141.5	84	102	122
Distribution	99	78	157	201
Distribution	63.5	107	107	100
Distribution	79	86	94	109
Distribution	73.5	65	92	142
Distribution	68	72	109	151
Distribution	61	70	100	142
Distribution	103	77	83	107
Distribution	82	72	99	137
Distribution	70	73	83	114
Distribution	50	88	150	170
Distribution	75	87	79	91
Distribution	50	84	136	162
Distribution	87	67	89	134
Distribution	92	55	83	147
Distribution	75	55	71	129
Distribution	86	58	81	140
Distribution	66	70	132	189
Distribution	69	74	91	123
Distribution	69	100	96	96
Distribution	59	98	110	112
Distribution	57	56	91	163
Distribution	84	58	98	167
Distribution	72	78	101	130
Distribution	84	98	111	113
Distribution	63	86	125	146
Distribution	71	89	84	94
Distribution	59	97	122	126
Distribution	80	75	64	85
Distribution	79	62	95	153
Distribution	78	81	60	73
Distribution	72	79	82	104
Distribution	65	85	109	129
Distribution	68	109	104	96
Distribution	72	78	92	106
Distribution	66	86	88	102
Distribution	66	103	100	97
Distribution	73	71	71	100
Distribution	58	72	88	121
Plant Effluent	89	107	75	70
Raw	77	NA	284	NA
Raw	77	NA	299	NA
Raw	113	118	106	90
Raw	96	147	78	53
Average	79.0	89.0	107.2	120.5

Abbreviations: DAL, double agar overlay method; SAL, single agar overlay method; NA, not analyzed.

Table 5.7
Recovery Efficiency for MS2 Phage

Sample	Actual PFU spiked (via DAL)	Sterile DI Water: % recovery based on DAL assay	Matrix spike: % recovery based on DAL assay	Matrix spike: % recovery based on SAL assay
DI water	4	150	NA	NA
Raw	57	NA	61	NA
Raw	57	NA	46	NA
Plant Effluent	108	39	36	93
Raw	108.5	31	42	135
Raw	172	111	162	146
Distribution	140.5	32	40	124
Distribution	109.5	46	26	57
DI water	130.5	46	NA	NA
Distribution	96	38	44	117
Distribution	120	39	29	74
Distribution	173.5	28	24	88
Distribution	114	36	28	78
Distribution	97.5	28	35	126
Distribution	119	26	34	132
Distribution	111	33	24	73
Distribution	82.5	24	25	105
Distribution	79.5	34	20	59
Distribution	98.5	25	19	72
Distribution	63	27	27	94
Distribution	65	19	20	108
Distribution	87	32	29	89
Distribution	65	17	23	136
Distribution	66	26	20	76
Distribution	46	43	33	75
Distribution	56	21	34	158
Distribution	44	34	32	93
Distribution	70	29	24	85
Distribution	79	28	19	68
Distribution	100	22	26	118
Distribution	91.5	31	20	64
Distribution	78	22	23	106
Distribution	75	37	28	75
Distribution	101	33	31	94
Distribution	134	31	40	132
Distribution	92	23	18	81
Distribution	99	16	24	150
Distribution	102	19	35	189
Distribution	78	19	17	87
Distribution	74	36	26	70
Distribution	78	19	37	193
Distribution	73	30	38	127
Distribution	101	20	30	150
Distribution	72	39	33	86
Distribution	105	33	25	74
Distribution	79	11	23	200
Distribution	50	48	72	150
Distribution	67	45	43	97
Distribution	83	33	24	77
Distribution	62	26	26	100
Distribution	76	45	33	94
Distribution	79	37	42	114
Distribution	82	35	38	107
Average	87.8	34.4	33.5	106.0

Abbreviations: DAL, double agar overlay method; SAL, single agar overlay method; NA, not analyzed.

DATABASE QA/QC

The database was examined for outliers and questionable data. One finished water *Bacillus subtilis* data point was flagged and removed from analysis. This data point was a misreported raw water value (the volume and method used for the samples was for raw water, not that for finished water).

Several questionable phage results were flagged where the negative control showed contamination. This contamination affected the data processed on 3/27/01, 4/10/01, and 4/17/01 and resulted in the flagging of 3 raw water and 15 distribution system samples (Table 5.8). Somatic coliphage were only detected in plant effluent and distribution system during this period, raising the likelihood that the results were invalid due to contamination. Although several male-specific coliphage samples were flagged during this period, removal of these data resulted in changes in the mean male-specific coliphage concentration for raw, plant effluent, and distribution system by only 0.07 to 0.11 pfu/100 mL (Table 5.8).

Table 5.8
Summary of database outliers and flagged data

Parameters	Monitoring Location	Contaminated Samples	N[a]	N < DL[b]	Mean	STD	Min	25%ile	50%ile	75%ile	90%ile	95%ile	Max
Bacillus subtilis	Effluent	with	75	32	192.24	1615.990	0.00	0.00	1.00	2.00	30.00	38.00	14000.00
cfu/100 mL		without	74	32	5.65	14.182	0.00	0.00	1.00	2.00	22.00	36.00	76.00
	Raw water	with	72	43	3.17	7.480	0.00	0.00	0.00	2.00	10.00	25.00	42.00
		without	69	65	3.10	7.556	0.00	0.00	0.00	2.00	10.00	25.00	42.00
Phage F-Amp	Effluent	with	72	66	0.21	0.871	0.00	0.00	0.00	0.00	0.00	1.00	5.00
(pfu/100 mL)		without	68	65	0.10	0.626	0.00	0.00	0.00	0.00	0.00	1.00	5.00
	Distribution system	with	408	376	0.21	0.942	0.00	0.00	0.00	0.00	0.00	1.00	10.00
		without	393	371	0.10	0.507	0.00	0.00	0.00	0.00	0.00	1.00	5.00
	Raw water	with	73	1	186.71	254.991	0.00	66.00	100.00	197.00	394.00	675.00	1645.00
		without	70	1	191.99	259.016	0.00	66.00	102.00	249.00	407.50	675.00	1645.00
Phage CN13	Effluent	with	73	71	0.08	0.595	0.00	0.00	0.00	0.00	0.00	0.00	5.00
(pfu/100 mL)		without	69	69	0.00	0.00	0.00	0.00	0.00	0.00	0.00	0.00	0.00
	Distribution system	with	414	410	0.21	3.657	0.00	0.00	0.00	0.00	0.00	0.00	74.00
		without	399	399	0.00	0.00	0.00	0.00	0.00	0.00	0.00	0.00	0.00

[a] N, number of samples
[b] N<DL, number of samples less than the detection limit (i.e., non-detect samples)

CHAPTER 6
RAW WATER RESULTS

As outlined in Chapter 3, the Upper Mississippi River is a major navigational waterway with multiple watershed uses and numerous sources of fecal pollution. One of the reasons for selecting the Davenport study site was because the source water was considered "microbially challenged." Water quality monitoring of the Mississippi River was conducted by collecting daily, weekly, or monthly samples at the river intake (Figure 6.1) using methods outlined in Chapter 4. The purpose of this chapter is to document the raw water quality results and compare these values to those collected in a previous AwwaRF study (LeChevallier et al., 2002).

Figure 6.1. Collection of grab samples at the raw water intake

PATHOGENS AND FECAL BACTERIA

Project monitoring began on October 30, 2000 and was completed on May 31, 2002. A summary of raw water data is presented in Table 6.1. *Giardia* cysts were detected 23 of 74 times (31.1%) (mean 0.093 cysts/L, minimum 0.0, maximum 1.1 cysts/L) using Method 1623 in the raw water with the majority of positive samples occurring during the November through April period (Figure 6.2). *Cryptosporidium* oocysts were detected 10 of 74 times (13.5%) (mean 0.019 oocysts/L, minimum 0.0, maximum 0.3 oocysts/L) using Method 1623 in the raw water with most of the positive samples occurring during the period of December through March of 2001, however the same trend was not observed in 2002 (Figure 6.3). This seasonal distribution was consistent with data previously collected at this site (LeChevallier et al., 2002). Two of 72 samples (2.8%) were positive for infectious *Cryptosporidium* using the CC-PCR method. These samples were positive in August 2001, and January 2002. Method USEPA 1623 did not simultaneously detect any oocysts in the CC-PCR-positive samples. The reasons for the lack of simultaneous results have previously been explained (LeChevallier et al., 2002), and can be attributed to splitting samples with low oocyst concentrations, and the low recovery efficiencies for the methods. Based on homologies to published sequences, analysis of the *hsp*70 amplicon showed that one isolate was C. *meleagridis,* while the other was the bovine genotype of *C. parvum*. Both species have been shown to cause infections in humans.

Table 6.1
Physiochemical and biological water quality parameters for the Mississippi River at Davenport, IA

Parameter	N[a]	N < DL[b]	Mean	STD	Min	25%ile	50%ile	75%ile	90%ile	95%ile	Max
pH (pH units)	577		8.3	0.4	7.6	8.0	8.2	8.4	8.8	9.0	9.3
Temperature (°C)	577		12.08	8.99	1.11	3.33	11.10	18.32	26.09	28.31	31.08
Turbidity (NTU)	577		24.1	21.7	2	9	20	33	45	60	197
Alkalinity (mg/L as $CaCO_3$)	577		156.3	18.8	95	152	161	169	175	177	183
Hardness (mg/L as $CaCO_3$)	381		186.3	20.0	120	181	190	199	206	210	226
Fecal coliform (cfu/100 mL)	392		341.8	589	3	48	113	316	969	1511	4480
Total coliform (cfu/100 mL)	420		4620	5960	100	1000	2088	5600	13396	18634	35000
Clostridium perfringens (cfu/100 mL)	75	0	41.6	56.3	9	16	25	46	70	149	410
Mean male specific coliphage (pfu/100 mL)[e,j]	69	42	3.1	7.6	0	0	0	2	10	25	42
Mean somatic coliphage (pfu/100 mL)[e,k]	70	1	192.0	259.0	0	66	102	249	408	675	1645
Culturable enteric viruses (MPN/100 L)	19	8	6.72	10.80	0.00	0.00	1.02	10.21	27.90	37.99	37.99
Bacillus subtilis (cfu/100 mL)[g]	74	0	16500	12900	1000	7000	13000	20000	36000	44000	62000
Giardia (cysts/L)[c,i]	74	51	0.093	0.202	0.000	0.000	0.000	0.100	0.300	0.400	1.100
Cryptosporidium (oocysts/L)[c,i]	74	64	0.019	0.054	0.000	0.000	0.000	0.000	0.100	0.100	0.300
Algae (cells/L)[l,i]	71		15800000	25200000	25700	1810000	6830000	17300000	32400000	86200000	130000000
Precipitation (in)	239		0.24	0.38	0.00	0.01	0.08	0.32	0.77	1.07	2.50
River flow (cfs)	577		64990	53990	18800	33300	42500	70500	125700	207900	275000
River stage (ft)	577		7.98	4.11	3.58	5.37	6.26	8.80	13.16	18.57	22.30

Number of Detects[c]				*Cryptosporidium*[d]		
	Giardia	N[a]	23		N[a,f,j]	72
	Cryptosporidium	N[a]	10		Positive[i]	2
					Negative	70

[a] Number of samples [b] Number of samples below the detection limit i.e. not detected. [c] Method 1623. [d] Method CC-PCR.
[e] Two samples were received by the laboratory but they were not analyzed. [f] Two failed PCR. [g] One sample lost due to lab accident.
[h] Positive by CC-PCR but not by USEPA Method 1623. [i] One sample was not collected. [j] Four samples were contaminated. [k] Three samples were contaminated. [l] Three samples were received by the laboratory but they were not analyzed.

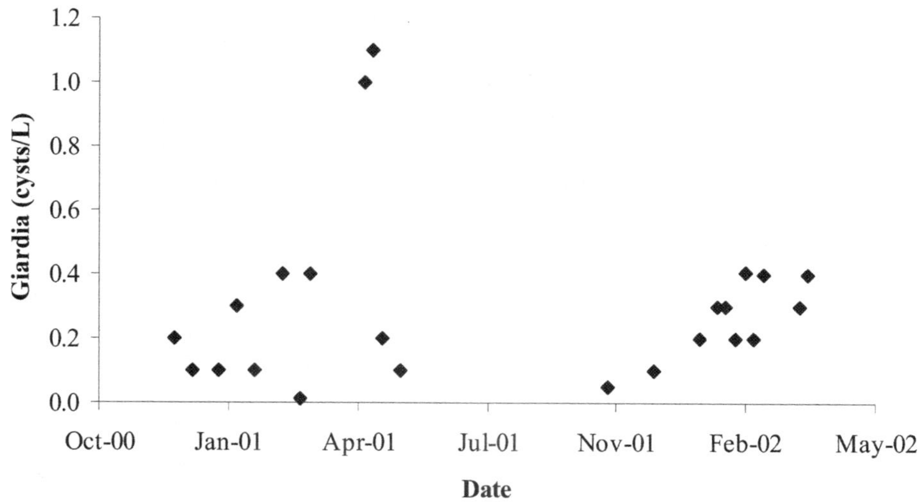

Figure 6.2. Distribution of *Giardia* cyst concentrations detected in the Mississippi River

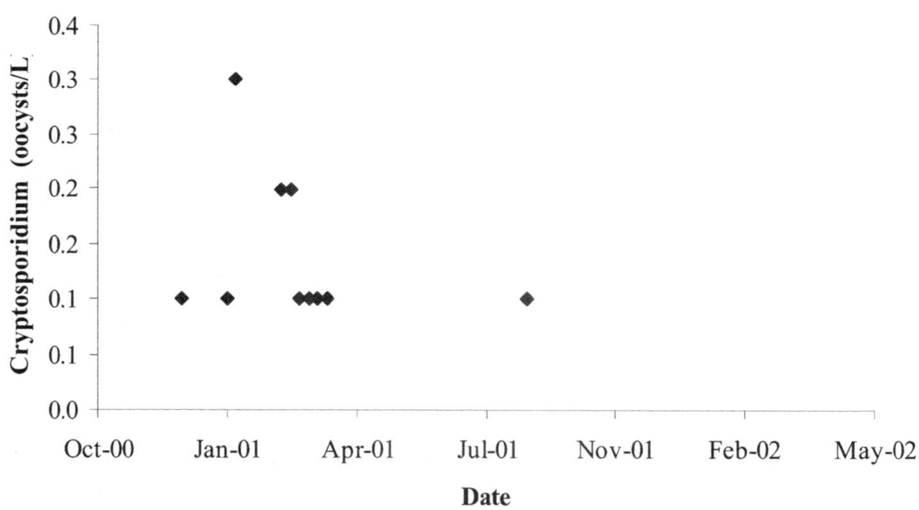

Figure 6.3. Distribution of *Cryptosporidium* oocyst concentrations detected in the Mississippi River

Culturable viruses were detected in 11 of 19 (57.9%) of the monthly raw water samples, with an average 6.72 MPN/100 L). Peak concentrations of viruses occurred in February of both 2001 and 2002 (Figure 6.4). Eight of the cell culture positive samples were examined using RT-PCR with primers for enteroviruses and reovirus. All samples were positive for reovirus, but negative for enterovirus. Four of the eight reovirus positive samples were subsequently cloned and sequenced. Sequence comparison using BLAST searches indicated 90% to 100% homology with other known reovirus sequences. None of the positive virus sequences matched the reovirus type 3 that was used as RT-PCR positive control, suggesting that the positive viruses were from natural occurrences and not due to laboratory contamination.

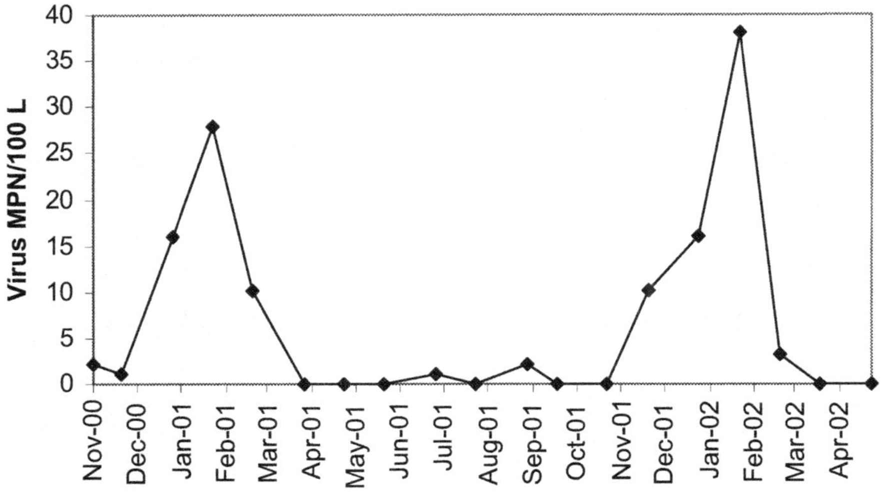

Figure 6.4. Distribution of virus concentrations detected in the Mississippi River

Fecal coliforms were detected in all of the 392 raw water samples tested. The mean fecal coliform level was 340 cfu/100 mL (minimum 3, maximum 4,480 cfu/100 mL). Likewise, total coliform bacteria were always detected in the raw water, with an average of 4,600 cfu/100 mL (minimum 100, maximum 35,000 cfu/100 mL). The continued presence of fecal indicator bacteria, protozoan pathogens, and enteric viruses clearly demonstrate the microbial health risks associated with untreated Mississippi River water.

INDICATOR ORGANISMS

Monitoring of spore-forming bacteria in raw water samples was performed to develop comparisons with filtered water results so that assessments can be made regarding treatment plant performance. In this regard, detection of *Bacillus* spores proved most useful, as all 74 samples analyzed were positive (Figure 6.5). The mean *Bacillus subtilis* spore concentration was 16,500 cfu/100mL, with a minimum level of 1,000 cfu/100mL and a maximum of 62,000 cfu/100mL. These raw water concentrations allow a 3 to 4.8 log assessment of treatment efficiencies. *Clostridium perfringens* spore levels were lower (mean 42 cfu/100 mL minimum 9 and maximum 410 cfu/100 mL). Although all 75 samples analyzed for *C. perfringens* spores were positive (Figure 6.5), the spore concentrations only permit a 1 to 2 log assessment of treatment efficiencies with further efforts to concentrate larger volumes of treated water.

The average mean concentration of male-specific coliphage was 3.1 pfu/100 mL (minimum 0, maximum 42 pfu/100 mL) and the viruses were detected in 27 of 69 (39.1%) of the raw water samples examined. Somatic coliphage had a mean of 192 pfu/100 mL (minimum 0, maximum 1,600 pfu/100 mL), with detection in over 98% of the samples examined (69 of 70 samples). Overall, male-specific coliphage were frequently detected January through May, 2001 and in the Fall of 2001, but the pattern did not continue into 2002 (Figure 6.6). Somatic coliphage were routinely detected with the lowest levels occurring during the summer of 2001.

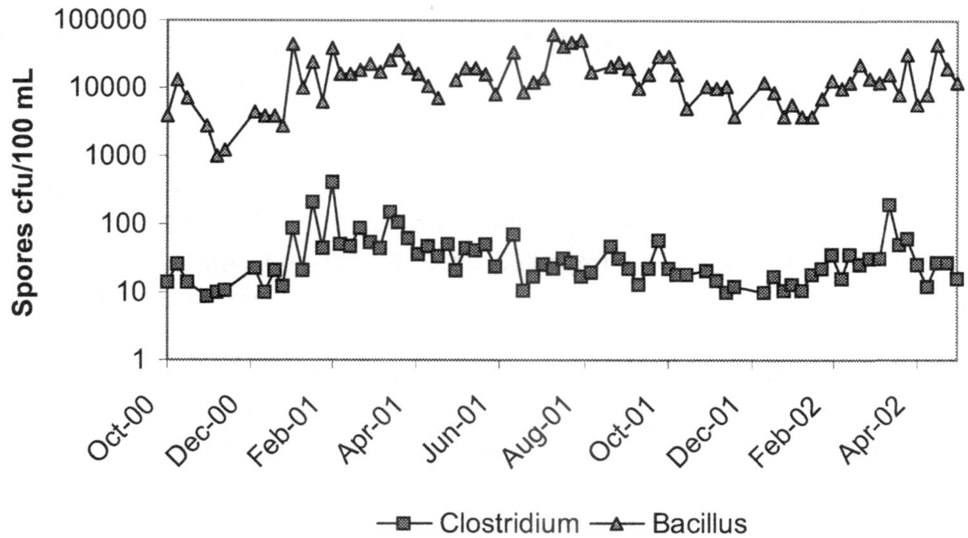

Figure 6.5. Raw water concentrations of *Clostridium perfringens* and *Bacillus subtilis* spores

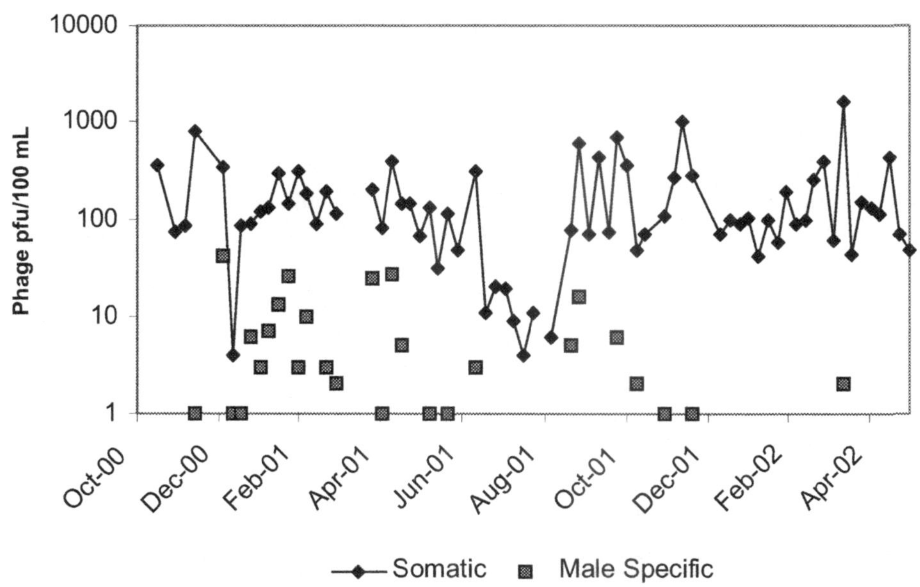

Figure 6.6. Raw water concentrations of somatic and male-specific coliphage

Figure 6.7 displays profiles for precipitation, river flow, and source water turbidity. The average precipitation was 0.24 inches, with the maximum rainfall of 2.50 inches in May 2002. A cluster of precipitation events in April and May 2001 was responsible for a peak river flow of 275,000 cubic feet per second (cfs) in April 2001, more than a four-fold increase over the average flow of 64,995 cfs. The profile of the river stage matched that of flow, with an average of 7.98 ft and a maximum height of 22.30 ft. These increases in flow were also associated with increases in turbidity. There were two events where the turbidity peaked above 100 NTU. The first was associated with the leading edge of the increase in river flow (e.g., first flush), and the second was related to a second flush event in May 2001. This second event resulted in a turbidity peak of 197 NTU. The average turbidity of the Mississippi River was 24.1 NTU.

The 2001 FLOOD EVENT

Figure 6.8 shows that the Mississippi River experienced a flood of record, peaking at 22.3 feet (flood stage, >15 ft) between April 14 and May 23, 2001. During this time a number of upstream wastewater facilities were impacted and raw sewage entered the river. Raw water quality parameters during the flood period are summarized in Table 6.2. The largest increases were observed for *Giardia* cysts, male-specific coliphage, total and fecal coliforms, and flow rate. During this period *Giardia* levels peaked at 1.0 cysts/L and total and fecal coliforms showed similar flood-related peaks (Figure 6.8), however, *Cryptosporidium* oocysts and culturable viruses were not detected in the floodwaters.

Table 6.2

Average Water Quality Parameters between April 14 and May 23, 2001

Parameter	N [a]	Average	% Difference [b]
pH units	40	7.89	-5
Temperature (°C)	40	16.5	37
Turbidity (NTU)	40	41.6	73
Alkalinity (mg/L as $CaC0_3$)	40	117	-25
Hardness (mg/L as $CaC0_3$)	28	158	15
Fecal Coliform (cfu/100 mL)	28	1440	320
Total Coliform (cfu/100 mL)	33	16,800	263
River Flow (ft^3/s)	40	224,700	246
River Stage (ft)	40	19.54	144
Precipitation (in)	22	0.425	75
Total *Giardia* counted (cysts/L)	6	0.4	330
Total Crypto per L (oocysts/L)	6	0	-100
Bacillus subtilis (cfu/100 mL)	5	18,000	9
Clostridium perfringens (cfu/100 mL)	6	60.2	45
Somatic coliphage (pfu/100 mL)	6	191	1
Male Specific coliphage (pfu/100 mL)	6	11.6	274
Culturable viruses (MPN/100 L)	1	0	-100

[a] Number of samples , b Difference from average values in Table 6.1.

Tim Wade, of the UC Berkeley epidemiology team, reported that highly credible gastrointestinal illness (HCGI) rates increased between April 14 and May 23 (Wade et al. 2002). For the Davenport area, based on 519 person-years of observations, the overall rate of HCGI was 2.52 episodes/person-yr and the overall rate of diarrhea was 0.75 episodes/person-yr. The rates were the highest in winter, November-February (HCGI: 3.28, diarrhea: 0.95). After controlling for season, the flood period was associated (p=0.006) with elevated levels of HCGI (rate ratio=1.23; 95% CI [1.04-1.47]). Diarrhea rates were also elevated during the flood period (rate ratio=1.14; 95% CI [0.97-1.34]), but of borderline statistical significance (p=0.074). Although it cannot be determined from this investigation exactly the source of the excess illness, adults with direct contact with floodwater (whose homes were flooded) were at increased risk of illness. Both adults and children whose homes or yards were flooded experienced high rates of HCGI. Children involved with flood cleanup activities experienced slightly higher rates of HCGI than those who did not. The stronger relationship between direct contact with floodwater and GI illness for children than adults is possibly due to increased susceptibility. For children under 5, any contact with floodwater was strongly associated with HCGI (IRR=3.18, 95% CI 1.79-5.66). The increased incidence of illness during the flood and the relationship between contact with floodwater and GI illness could not be explained by other available covariates.

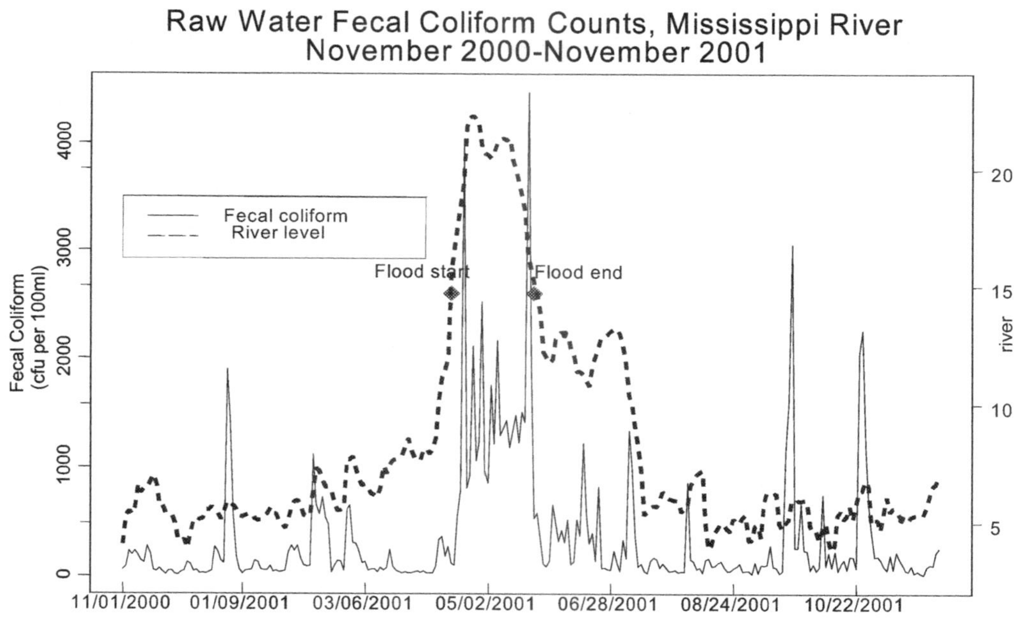

Figure 6.7. Relationship between fecal coliform level and flood stage, Spring 2001

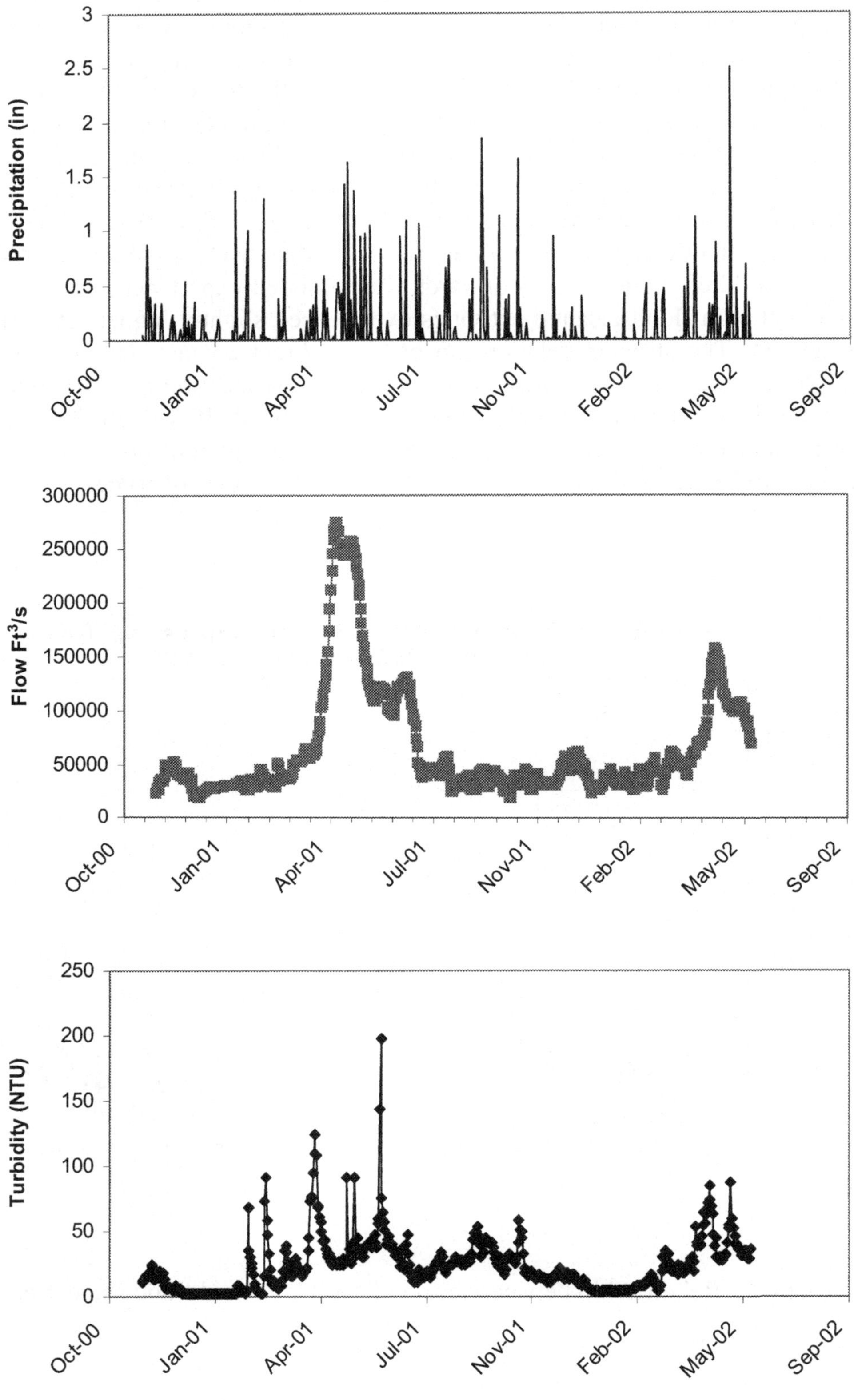

Figure 6.8. Precipitation, flow and turbidity at Davenport, IA

COMPARISON TO PREVIOUS DATA

The AwwaRF project "Source Water Assessment: Variability of Pathogen Concentrations" (#488) (LeChevallier et al. 2002) conducted extensive analysis of the occurrence of *Giardia* cysts, *Cryptosporidium* oocysts and indicator organisms in six watersheds, including the Mississippi River at Davenport from June 1999 to June 2000. It was of interest to compare the database of AwwaRF 488 to the data from this study (AwwaRF 2580) to examine the year-to-year variability of the water quality parameters. These comparisons are presented in Figures 6.9 to 6.23. A visual comparison between both studies showed similar trends for the majority of measured parameters. No comparisons were provided for *Bacillus subtilis* spores or culturable viruses because these parameters were not collected during the AwwaRF 488 project.

Descriptive statistics including sample size, mean, standard deviation, minimum, maximum, and percentiles (50%, 90%, 95%) for all biological and physiochemical parameters are shown in Tables 6.3 and 6.4. To compare the data from the two surveys, it was necessary to determine if the environmental variables followed normal distributions. The Shapiro-Wilk test was used since the datasets included less than 2,000 observations. Most of the AwwaRF 2580 parameters diverged from a normal distribution at the 95% confidence level ($p < 0.05$). These results were similar to those obtained in AwwaRF 488 where only hardness and alkalinity had normal distributions (LeChevallier et al. 2002). The non-parametric statistical Wilcoxon Rank Sum method compared the concentrations of biological, physical, and chemical environmental factors measured in the two projects. As shown in Table 6.5 the majority of the data collected in the two surveys were not significantly different at the 95% confidence level ($p \geq 0.05$), but temperature, fecal coliform, *Giardia* cysts, and somatic coliphage were statistically different between the two studies ($p < 0.05$).

Cryptosporidium oocysts levels averaged 0.017 and 0.019 oocysts/L for the AwwaRF 488 and 2580 projects, respectively. These results demonstrate a relative consistency in *Cryptosporidium* occurrence, and support the conclusions of the AwwaRF 488 project that oocyst variability would be considerably less in a major water supply like the Mississippi River (LeChevallier et al. 2002). However, when a rolling 12-value average was calculated (to simulate a 12-month rolling average), there was a 20.3% likelihood of exceeding the 0.075 oocyst/L bin threshold as required by the LT2ESWTR (Figure 6.24). These data are slightly less than the 38% likelihood of exceeding the 0.075 bin threshold determined using the AwwaRF488 data, which were probably influenced by higher peak values (Table 6.3). The results emphasize the benefits of basing source water concentrations on a large set of data collected over a prolonged interval. Short-term averages, as required by the LT2ESWTR when only 24 monthly samples are collected, exaggerate the variability of source water oocyst concentrations.

Giardia cyst values, however, were significantly higher in AwwaRF 2580 compared to project 488 (Table 6.5). This finding was not attributed to the peak levels observed in the 2001 flood because when these data were analyzed without the flood results, *Giardia* levels were still higher than the AwwaRF 488 results. Data in Figure 6.11 demonstrate the fecal coliform results were elevated at other times during project 2580 and may have been related to Fall and Spring precipitation events (Figure 6.17). Although the average rainfall observed during AwwaRF 2580 was 2.4 times higher than that recorded during project 488, the difference was not significant.

The significant difference in temperature values between the two projects (Table 6.5) can be explained by the fact that AwwaRF 2580 extended over two winter periods, whereas project 488 covered only one seasonal cycle (Figure 6.22).

By comparison of the two project databases, it can be concluded that the source water quality parameters measured were fairly representative of the Mississippi River at Davenport, Iowa. The combination of the two datasets provides one of the most comprehensive databases available for characterizing raw water quality.

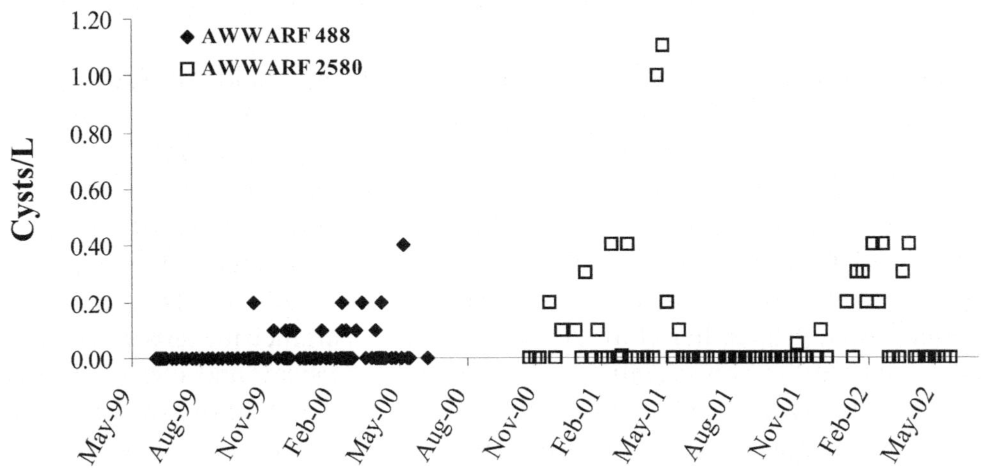

Figure 6.9. Comparison of *Giardia* cyst levels between AwwaRF projects 488 and 2580.

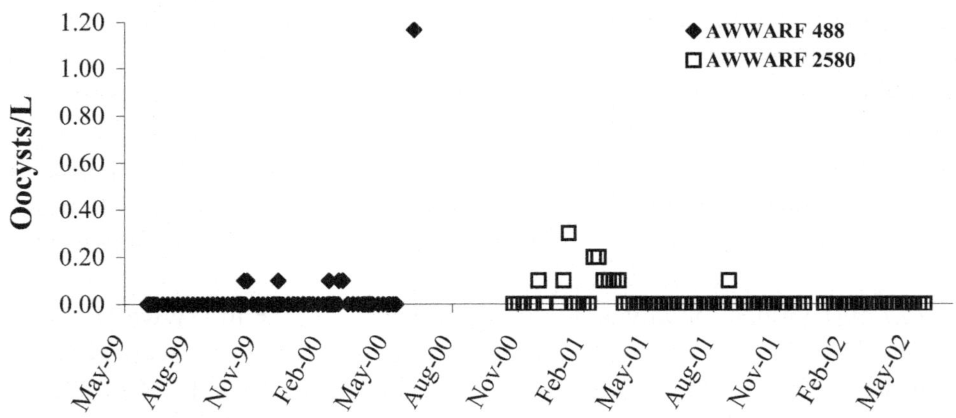

Figure 6.10. Comparison of *Cryptosporidium* oocyst levels between AwwaRF projects 488 and 2580.

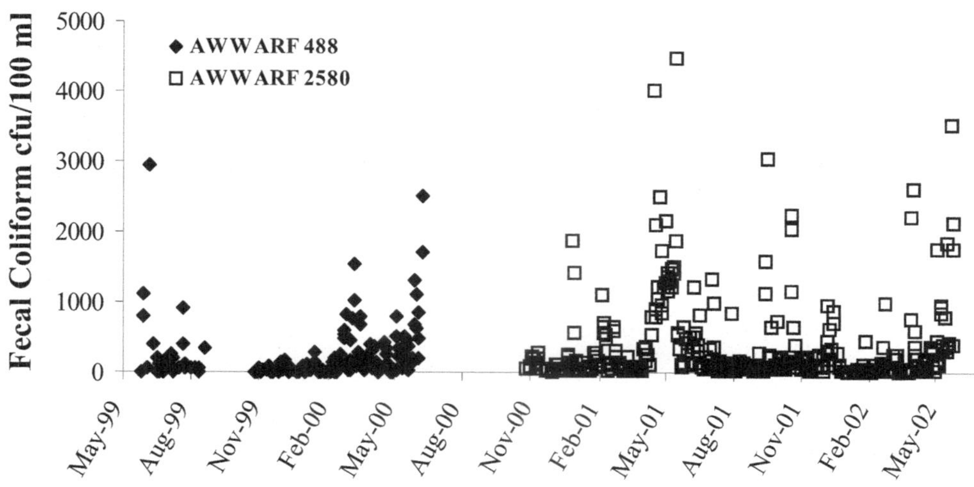

Figure 6.11. Comparison of fecal coliform levels between AwwaRF projects 488 and 2580.

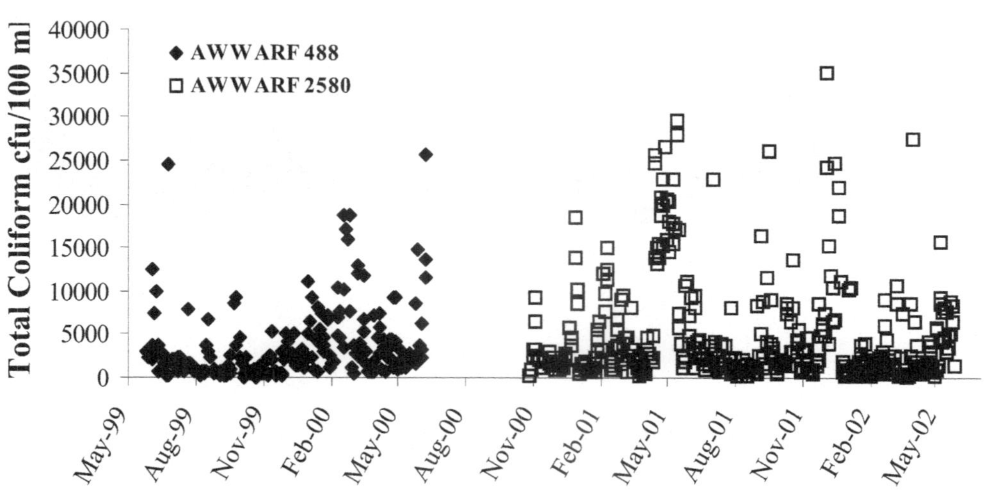

Figure 6.12. Comparison of total coliform levels between AwwaRF projects 488 and 2580.

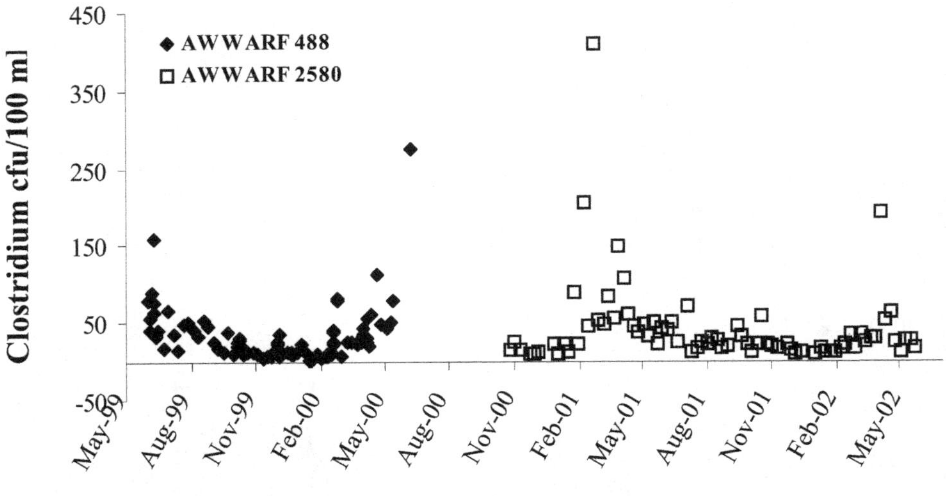

Figure 6.13. Comparison of *Clostridium perfringens* spore levels between AwwaRF projects 488 and 2580.

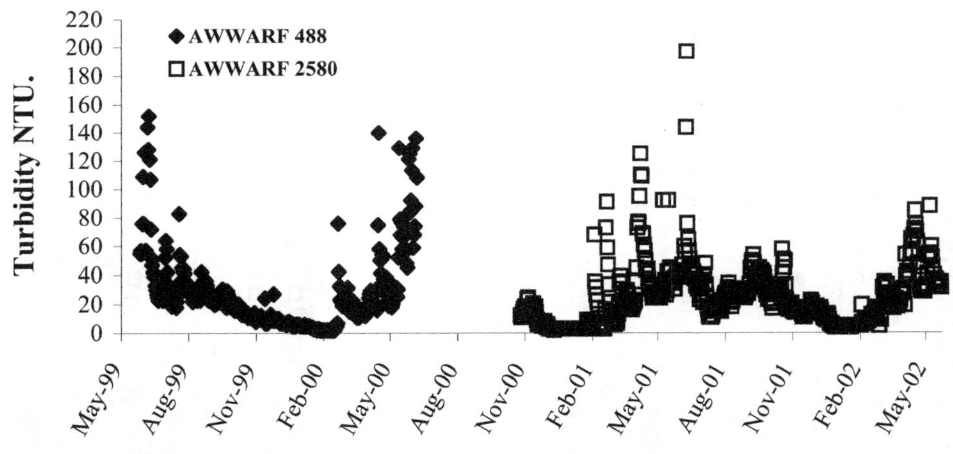

Figure 6.14. Comparison of turbidity levels between AwwaRF projects 488 and 2580.

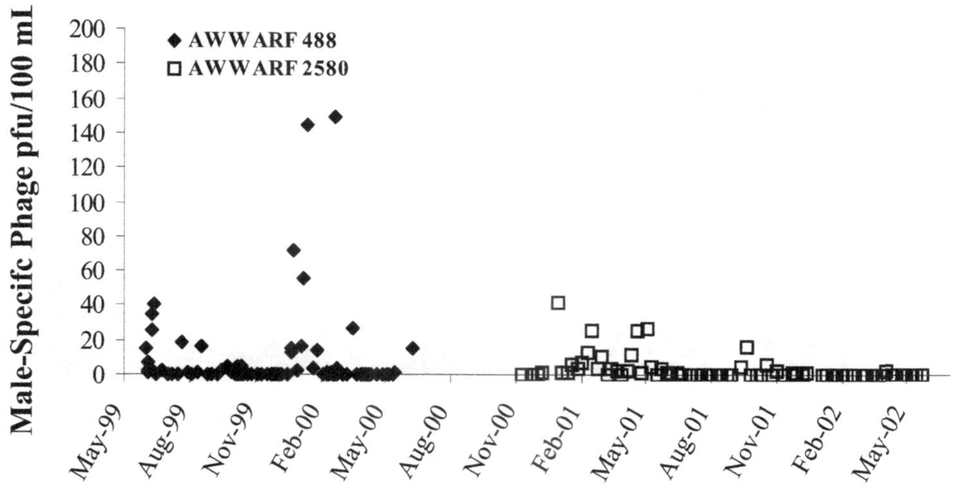

Figure 6.15. Comparison of male-specific coliphage levels between AwwaRF projects 488 and 2580.

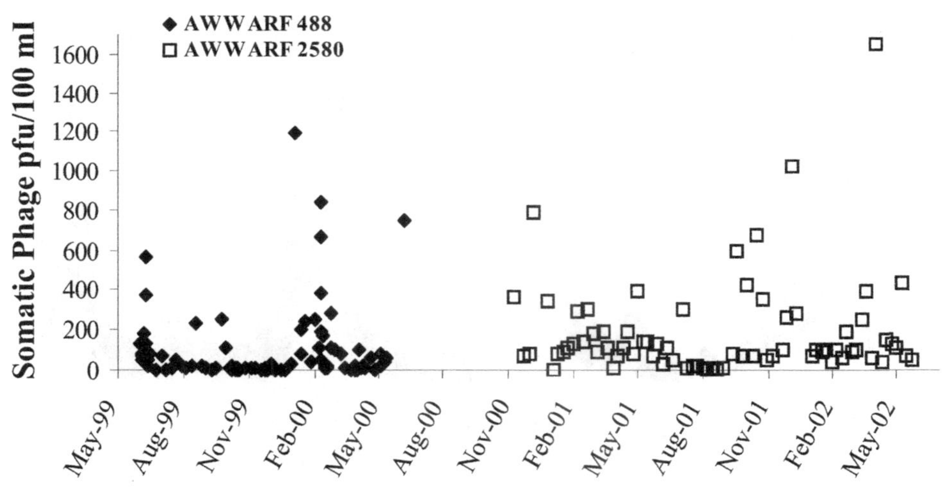

Figure 6.16. Comparison of somatic coliphage levels between AwwaRF projects 488 and 2580.

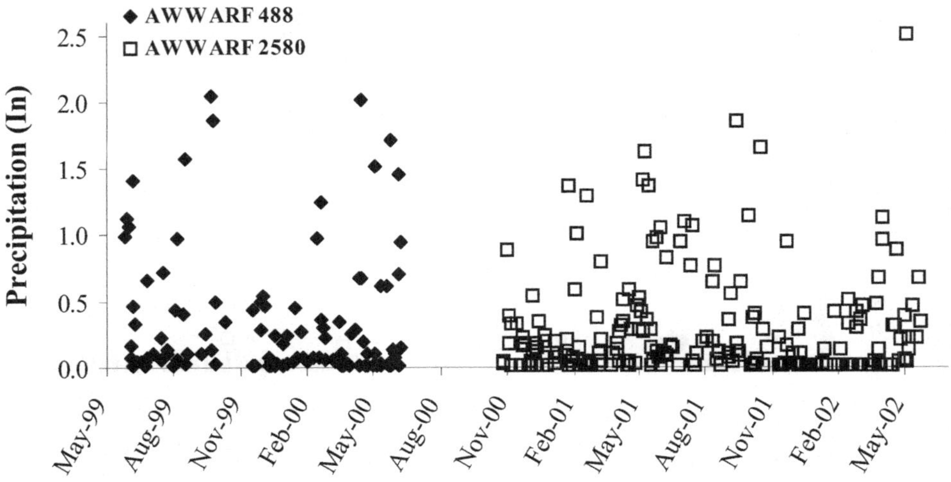

Figure 6.17. Comparison of precipitation levels between AwwaRF projects 488 and 2580.

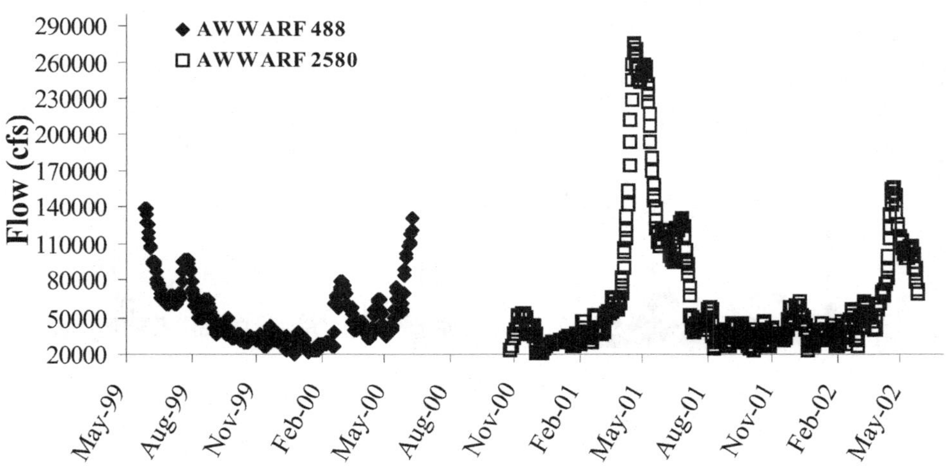

Figure 6.18. Comparison of river flow rates between AwwaRF projects 488 and 2580.

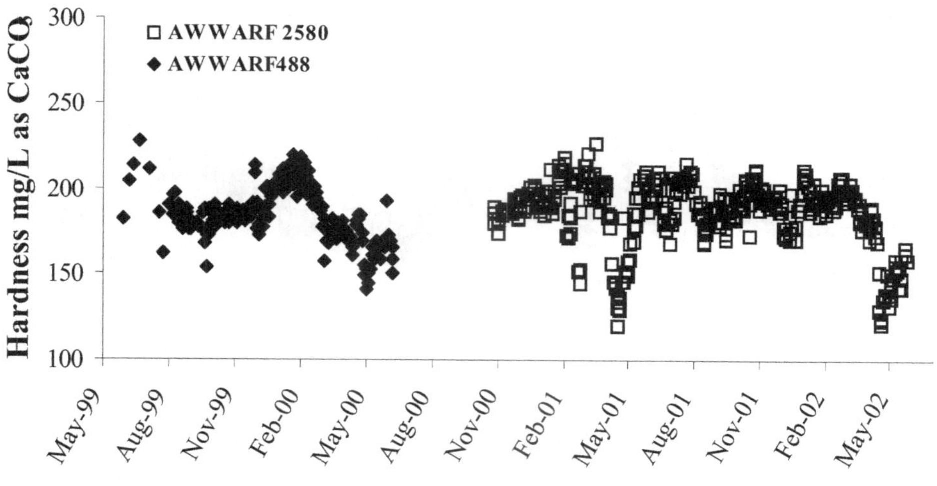

Figure 6.19. Comparison of hardness levels between AwwaRF projects 488 and 2580.

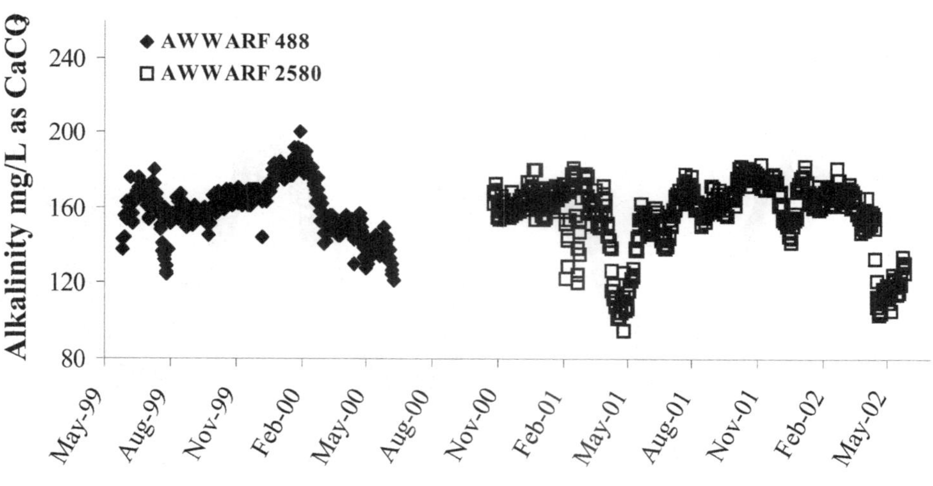

Figure 6.20. Comparison of alkalinity levels between AwwaRF projects 488 and 2580.

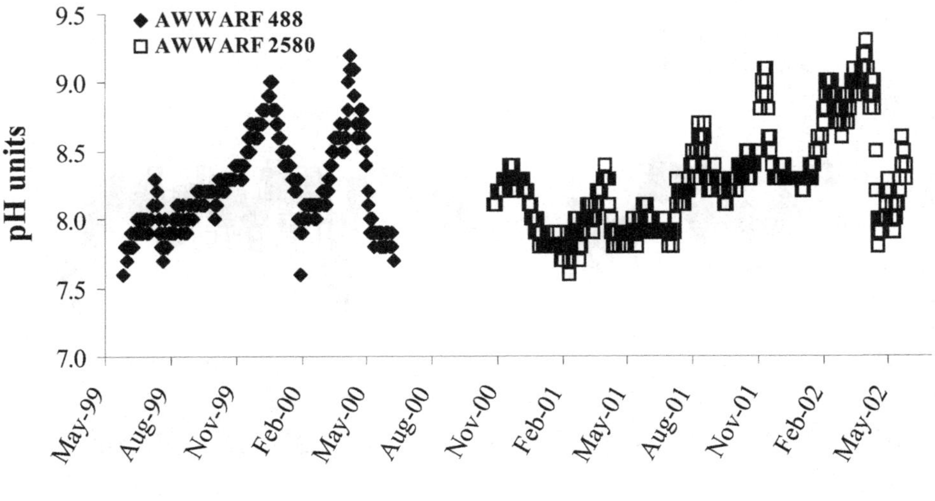

Figure 6.21. Comparison of raw water pH levels between AwwaRF projects 488 and 2580.

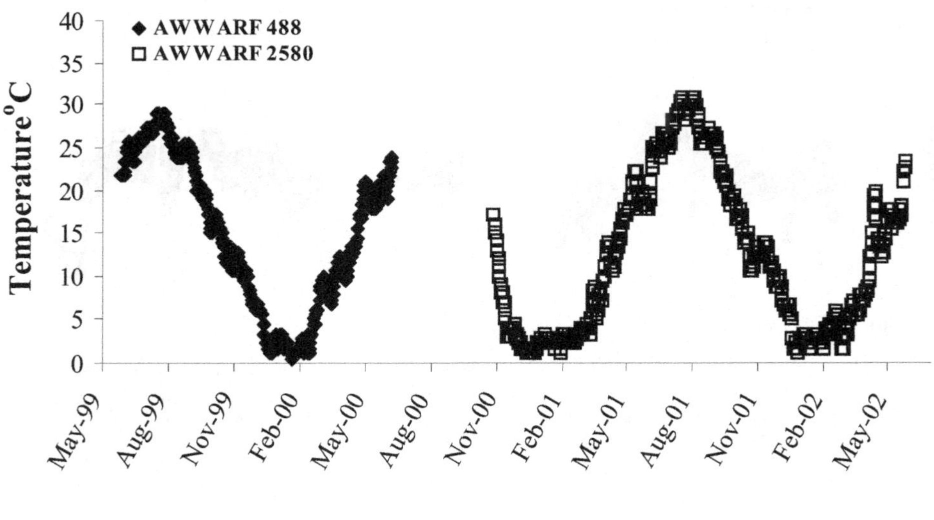

Figure 6.22. Comparison of raw water temperature levels between AwwaRF projects 488 and 2580.

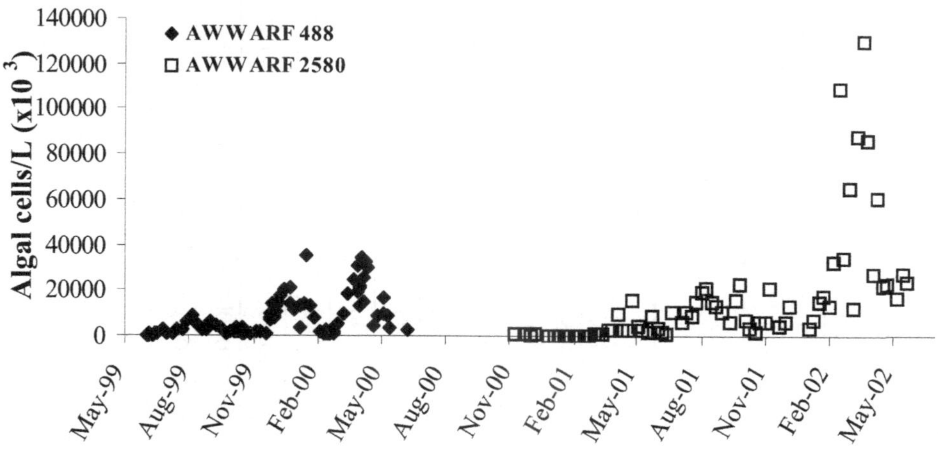

Figure 6.23. Comparison of algal cell levels between AwwaRF projects 488 and 2580.

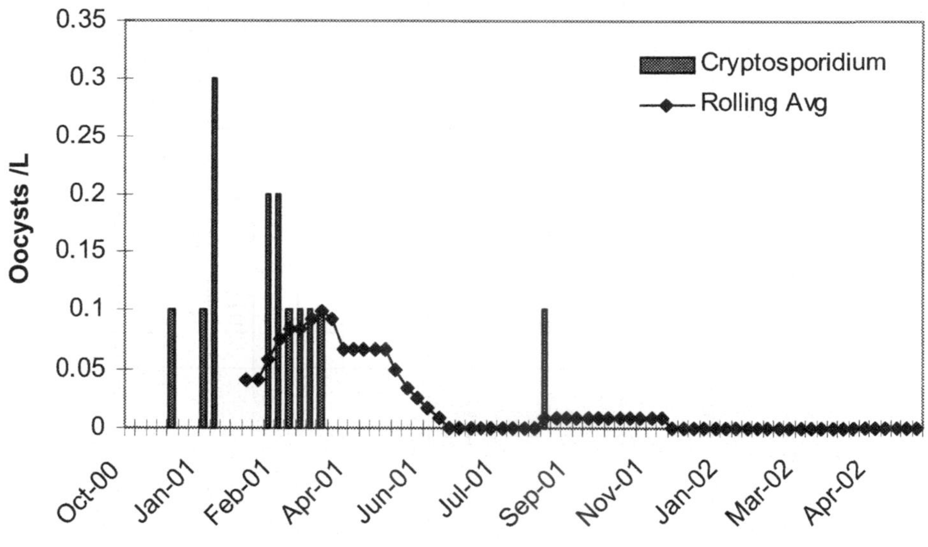

Figure 6.24. Calculation of a rolling 12-value average to simulate the LT2ESWTR monitoring requirements.

Table 6.3
Summary statistics for biological water quality parameters collected during AWWARF Projects 2580 and 488

Parameter	AwwaRF Project	N[A]	Mean	STD	Min	50%ile	90%ile	95%ile	Max
Fecal coliforms (cfu/100 mL)	2580	392	342	589	3	113	969	1510	4480
	488	186	247	632	0	60	622	900	6860
Total coliforms (cfu/100 mL)	2580	420	4620	5964	100	2090	13400	18600	35100
	488	222	3610	4140	0	2400	8600	12000	25600
Mean male-specific coliphage (pfu/100 mL)	2580	69	3.1	7.6	0	0	10	25	42
	488	101	11.7	49.5	0	0	16	40	449
Mean somatic coliphage (pfu/100 mL)	2580	70	192	259	0	102	408	675	1645
	488	101	126	388	3	16	251	563	3500
Clostridium perfringens spores (cfu/100 mL)	2580	75	41.6	56.3	9	25	70	149	410
	488	101	31.9	35.6	2	23	64	80	274
Giardia[B] (cysts/L)	2580	74	0.093	0.202	0.000	0.000	0.300	0.400	1.100
	488	101	0.022	0.061	0.000	0.000	0.000	0.100	0.399
Cryptosporidium[B] (oocysts/L)	2580	74	0.019	0.054	0.000	0.000	0.100	0.100	0.300
	488	101	0.017	0.118	0.000	0.000	0.000	0.100	1.167
Algae (cells/L)	2580	71	15850000	25230000	25700	6830000	32400000	86200000	130000000
	488	101	7890000	9071000	197000	3141000	21500000	29780000	35030000

[A] Number of samples [B] USEPA Method 1623

Table 6.4
Summary statistics for physical and chemical water quality parameters collected during AwwaRF Projects 2580 and 488

Parameter	AwwaRF Project	N[a]	Mean	STD	Min	50%ile	90%ile	95%ile	Max
pH (pH units)	2580	577	8.3	0.4	7.6	8.2	8.8	9.0	9.3
	488	384	8.1	0.4	7.6	8.2	8.7	8.8	9.2
Turbidity (ntu)	2580	577	24.1	21.7	2	20	45	60	197
	488	384	33.8	105.8	2	20	58	84	1850
Temperature (°C)	2580	577	12.1	9.0	1.1	11.1	26.1	28.3	31.1
	488	384	14.6	8.8	0.6	14.2	26.1	27.2	28.9
Hardness (mg/L as $CaCO_3$)	2580	381	186	20.0	120	190	206	210	226
	488	222	184	15.8	141	183	208	211	228
Alkalinity (mg/L as $CaCO_3$)	2580	577	156	18.8	95	161	175	177	183
	488	383	159	14.5	122	161	180	182	200
Precipitation (inches)	2580	239	0.24	0.38	0.00	0.08	0.77	1.07	2.50
	488	379	0.10	0.30	0.00	0.00	0.28	0.67	2.05
River flow (cfs)	2580	577	65000	53990	18800	42500	125700	207900	275000
	488	384	50100	24900	6697	41200	87200	98100	139000

Table 6.5
Wilcoxon Rank Sum test p-values comparing AwwaRF Project 2580 and 488 datasets

Group	Parameter	P-value	Significantly Different?
Physiochemical	pH	0.39	No
	Temperature	<0.01	Yes
	Turbidity	0.77	No
	Hardness	0.06	No
	Flow	0.06	No
	Precipitation	0.59	No
Biological	Fecal coliforms	<0.01	Yes
	Total coliforms	0.31	No
	Giardia cysts	0.01	Yes
	Cryptosporidium oocysts	0.14	No
	Mean male specific coliphage	0.47	No
	Mean somatic coliphage	<0.01	Yes
	Algae	0.13	No
	Clostridium perfringens spores	0.07	No

CHAPTER 7
TREATMENT PLANT RESULTS

MICROBIAL DATA

Extensive monitoring of the East River Station in Davenport, IA was conducted to document the plant performance for control of microbial pathogens and the production of consistent water quality. A total of 577 daily samples were recorded for physiochemical parameters including pH, turbidity, alkalinity, total chlorine, particle count, phosphate (576), fluoride (516), and hardness (86). A summary of plant effluent data is shown in Table 7.1.

The treatment plant was uniformly negative for all measurements of infectious *Cryptosporidium* (71 100-L samples), culturable enteric viruses (19 1,000-L samples), total coliforms (408 100-mL samples), somatic coliphage (69 100 mL samples), and *C. perfringens* spores (75 100-mL) samples. These data indicate that no organisms of public health significance were detected in the treated water. *B. subtilis* spores were detected in 42 of 74 samples, with an average of 5.7 cfu/100 mL (minimum 0, maximum 76 cfu/100 mL). Comparison of paired *B. subtilis* spore levels in the raw and treated water showed an average removal of 3.5 logs (range 1.7 to 4.8 logs) (Table 7.2). Spores were typically low or absent for most of the year with the exception of September 2001 through January 2002 (Figure 7.1). It is possible that the increase in spore counts was associated with the introduction of the new 2.47 MG clearwell. However, the plant performance for turbidity and particle count removal was consistently good during this time.

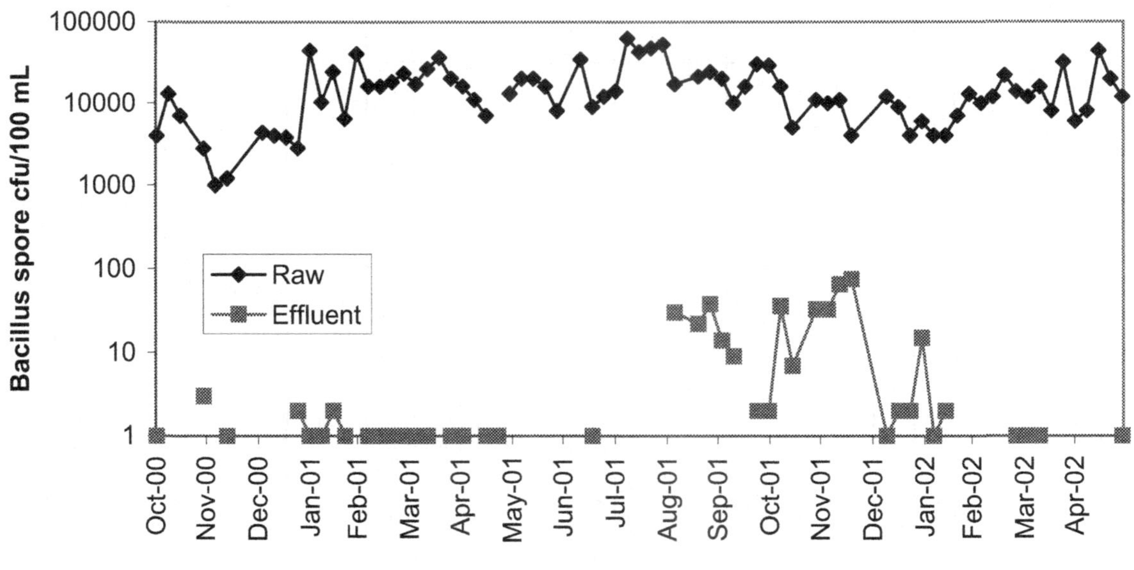

Figure 7.1. *Bacillus subtilis* spores in raw and treated water.

Table 7.1

Physiochemical and biological water quality parameters for the East River Station effluent, Davenport, IA

Parameter	N[a]	N < DL[b]	Mean	STD	Min	25%ile	50%ile	75%ile	90%ile	95%ile	Max
pH (pH units)	577		7.4	0.1	7.3	7.4	7.4	7.4	7.4	7.4	7.6
Turbidity (ntu)	577		0.052	0.01	0.03	0.05	0.05	0.06	0.07	0.07	0.09
Alkalinity (mg/L as $CaCO_3$)	577		138	16.1	85	132	141	148	154	157	164
Hardness (mg/L as $CaCO_3$)	86		186	18.7	129	183	190	197	203	206	212
Phosphate (mg/L)	576		1.08	0.08	0.86	1.01	1.08	1.15	1.19	1.20	1.26
Total chlorine (mg/L)	577		2.54	0.17	2.14	2.43	2.52	2.67	2.77	2.81	2.94
Fluoride (mg/L)	516		1.00	0.04	0.77	0.97	1.00	1.02	1.04	1.07	1.23
Total coliform (cfu/100 mL)	408	408	0.0	0.0	0	0	0	0	0	0	0
HPC (cfu/mL)	376	112	55.7	131.6	0	0	20	45	140	260	1430
Bacillus subtilis (cfu/100 mL)[g]	74	32	5.7	14.2	0	0	1	2	22	36	76
Clostridium perfringens (cfu/100 mL)	75	75	0.0	0.0	0	0	0	0	0	0	0
Culturable enteric viruses (MPN/100 L)	19	19	0.00	0.000	0.00	0.00	0.00	0.00	0.00	0.00	0.00
Mean male specific coliphage (pfu/100 mL)[d]	68	65	0.1	0.6	0	0	0	0	0	1	5
Mean somatic coliphage (pfu/100 mL)[e]	69	69	0.0	0.0	0	0	0	0	0	0	0
Water flow (mgd)	577		17.5	2.1	13.0	16.3	16.9	17.8	20.4	22.6	27.0
CT obtained/CT required	577		6.96	6.14	1.01	2.28	3.71	10.57	17.59	19.76	29.39
Particle count (#/mL)	577		78.4	103	3.42	20.7	43.7	101	180	268	1510

Cryptosporidium[c]	N[a,f]	71
	Positive	0
	Negative	71

[a] Number of samples. [b] Number of samples below the detection limit i.e. not detected. [c] Method CC-PCR. [d] Two samples were received by the laboratory but they were not analyzed. Five samples were contaminated. [e] Two samples were received by the laboratory but they were not analyzed. Three samples were contaminated and excluded from the analysis. [f] Three failed PCR. One sample was not collected. [g] One sample was ruled out as an outlier and excluded from the data analysis.

Table 7.2
East River Treatment Plant removal efficiencies [1]

Parameter	Mean Raw Water Concentration	Mean Effluent Concentration	% Removal	Log removal
Turbidity (ntu)	24.1	0.05	99.79	2.7
Particle count (number /mL)	2640*	78.41	97.0	1.5
Total coliform (cfu/100 mL)	4620	<0.002	100	>6.36
Bacillus subtilis (cfu/100 mL)	16500	5.7	99.97	3.5
Clostridium perfringens (cfu/100 mL)	41.6	<0.01	100	>3.61
Culturable viruses (MPN/100 L)	6.72	0	100	>2.12
Male specific coliphage (pfu/100 mL)	3.1	0.1	96.77	1.5
Somatic coliphage (pfu/100 mL)	192.0	<0.01	100	>4.1

[1] Calculation of removal was based on paired samples between raw and plant effluent samples.
* Particle count measurements were performed on the settled water (combined Superpulsator) and after the filters

The calculation of the log removal for male-specific coliphage was limited due to low raw water levels, but the removal of somatic coliphage was >4.1 logs through the treatment process (Table 7.2). Coliphage were detected in treated plant effluent water on three occasions. Three of 68 samples tested positive for male specific coliphage, with an overall average of 0.1 pfu/100 mL (minimum 0, maximum 5 pfu/100 mL). Table 7.3 shows that the coliphage results were scattered throughout 2001, with low level results in February, March, and September.

Table 7.3
Plant effluent coliphage results

Phage Type	Sample Date	Total volume assayed (L)	PFU/100 mL
Male Specific	2/6/01	100	1
Male Specific	3/13/01	100	5
Male Specific	9/25/01	100	1

PFU, plaque forming units.

Heterotrophic plate count bacteria (HPC) were enumerated using R2A medium incubated at 22°C (room temperature) for 7 days. This medium typically produces bacterial counts that are 10 to 1,000 times higher than the standard plate count medium (Reasoner and Geldreich 1985). HPC bacteria were detected in 264 of 376 samples analyzed, with a mean of 55.7 cfu/mL (minimum 0, maximum 1,430 cfu/mL). There was no obvious pattern to the occurrence of HPC bacteria in the treated plant effluent (Figure 7.2).

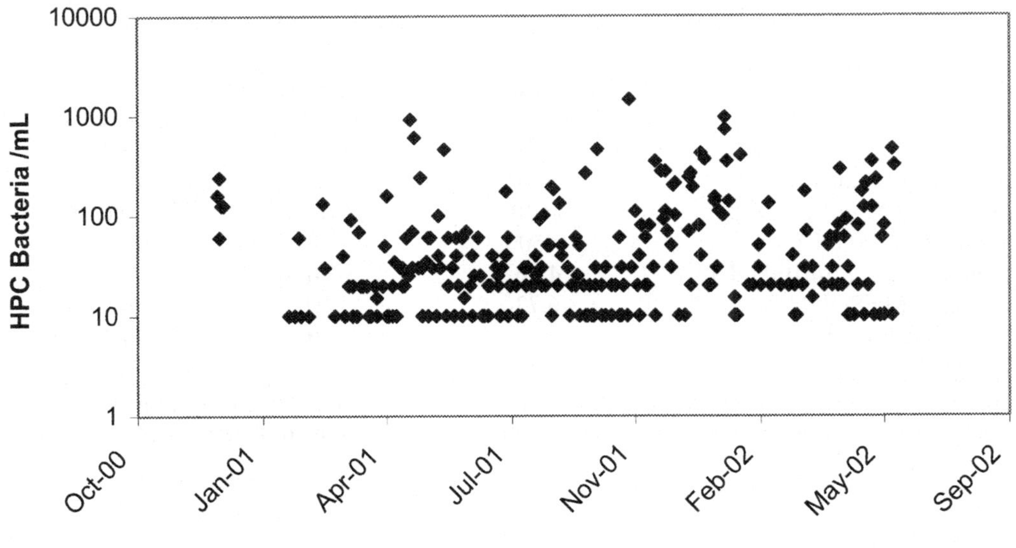

Figure 7.2. Plant effluent HPC bacteria.
Note that zero values are not shown on the log scale.

PHYSICAL AND CHEMICAL DATA

Sedimentation and Clarification

Monitoring of the two Superpulsator and the two sedimentation tanks was performed to document the performance of these processes for particle counts, flow, turbidity, alkalinity, pH, and free chlorine. Results are shown in Table 7.4. Particle count (>3 μm) data were collected only for the Superpulsator effluent and averaged 2,640 particles/mL (minimum 142, maximum 11,500 particles/mL). Concurrent raw water particle count data were not collected because the large source water values exceeded the acceptable coincidence range of the sensor.

The average turbidity for the two Superpulsators were 1.25 and 1.43 NTU (minimum 0.40, maximum 3.80 NTU). Similarly, the average turbidity for the two sedimentation tanks were 1.67 and 1.63 NTU (minimum 0.60, maximum 5.40 NTU). The Superpulsators tended to perform better than the sedimentation basins for turbidity removal, with 90 to 95% of the values [2.0 NTU (Table 7.4). Conversely, the sedimentation basins had 75% of the turbidity values [2.1 NTU. The total plant flow tended to be split evenly between the Superpulsator and sedimentation basin processes. Other physicochemical parameters, including alkalinity, pH, and free chlorine were similar between the two processes (Table 7.4). Wash water for backwashing the filters was collected from the clearwell and the average wash water flow was 0.4 MGD, but varied (range 0.13 to 2.08 MGD) depending on the number of filters washed. Spent filter wash water (and sludge) was discharged to a U. S. Army Corps of Engineers interceptor storm sewer.

Table 7.4
Physiochemical water quality data for Sedimentation and Clarification

Parameter	Sampling Location	N[a]	Mean	STD	Min	25%ile	50%ile	75%ile	90%ile	95%ile	Max
Particle count (#/mL)	Superpulsator[b]	572	2640	1730	142	1310	2170	3520	5230	6380	11500
Turbidity	Superpulsator 1[b]	571	1.43	0.48	0.40	1.10	1.40	1.70	2.00	2.30	3.80
	Superpulsator 2[b]	571	1.25	0.42	0.40	1.00	1.20	1.40	1.70	2.00	3.70
	Sedimentation Tank 1	569	1.67	0.63	0.70	1.20	1.50	2.10	2.50	2.70	5.40
	Sedimentation Tank 2	574	1.63	0.59	0.60	1.20	1.40	2.10	2.50	2.70	4.00
Alkalinity (mg/L as CaCO3)	Sedimentation Tank 1	569	129	21.1	68	122	136	145	150	152	161
	Sedimentation Tank 2	574	130	21.3	69	123	136	145	151	153	170
pH (pH units)	Superpulsator 1	571	7.4	0.21	7.0	7.3	7.4	7.5	7.7	7.8	8.1
	Superpulsator 2	571	7.4	0.20	7.0	7.3	7.4	7.5	7.7	7.8	8.1
	Sedimentation Tank 1	569	7.3	0.2	6.7	7.1	7.3	7.4	7.5	7.6	7.7
	Sedimentation Tank 2	574	7.3	0.2	6.8	7.1	7.3	7.4	7.5	7.6	7.7
Free Chlorine (mg/L)	Superpulsator	374	1.69	0.98	0.21	0.84	1.18	2.70	3.05	3.16	4.13
	Sedimentation Tank 1	537	1.41	0.69	0.30	0.90	1.13	1.87	2.47	2.68	3.37
	Sedimentation Tank 2	552	1.44	0.72	0.25	0.91	1.16	1.90	2.65	2.88	3.70
Flow (mgd)	Plant influent	577	18.09	2.09	13.15	16.86	17.54	18.50	21.00	22.97	26.20
	Superpulsator	573	8.75	1.26	1.23	8.00	8.60	9.57	10.23	10.75	12.10
	Sedimentation tank 1	575	4.65	1.10	0.43	4.13	4.54	4.99	5.82	6.53	10.84
	Sedimentation tank 2	576	4.81	1.30	0.96	4.095	4.56	5.09	6.37	7.62	11.62
	Washwater	577	0.4	0.2	0.13	0.32	0.39	0.49	0.6	0.69	2.08

[a] Number of samples [b]Data shown for "Superpulsator" are from the combined sample tap in the laboratory which yields a composite sample from both Superpulsators. Results of grab samples from individual Superpulsators are recorded as "Superpulsator 1" and "Superpulsator 2."

Filtration

The twenty dual media, granular activated carbon (GAC) and sand filters were monitored for turbidity, headloss, flow, and other operational parameters. Although filter effluent turbidities were measured continuously, plant operators log hourly values, and these hourly logs were averaged to produce a daily value. The mean plant effluent turbidity, based on 577 daily values, was 0.05 NTU, with a range between 0.03 and 0.09 (Table 7.1). Contrary to the *Bacillus* data, turbidity values tended to be low from September 2001 through January 2002 (Figure 7.3). Comparison of the individual filter effluent turbidity values to the plant effluent value using the non-parametric Wilcoxon Rank Sum test showed that all the filters were statistically lower (at the 95% confidence interval) than the effluent except for filter 11. A summary of the turbidity data for the 20 individual filter effluents is shown in Table 7.5. The mean filter turbidity value ranged from 0.034 to 0.055 NTU. Average daily turbidity values recorded for all twenty filters during the study are shown in Appendix F. Filter effluent turbidities were measured continuously and hourly values were recorded, and were used to calculate a daily average. Overall, for the 10,991 daily turbidity values recorded, 80.66% were less than 0.05 NTU, 19.15% were between 0.05 and 0.10 NTU, and 0.13% were between 0.10 and 0.15 NTU. Only six individual daily average filter measurements (0.06%) were greater than 0.15 NTU, all of which occurred between November 14 and December 20, 2000 (Table 7.6).

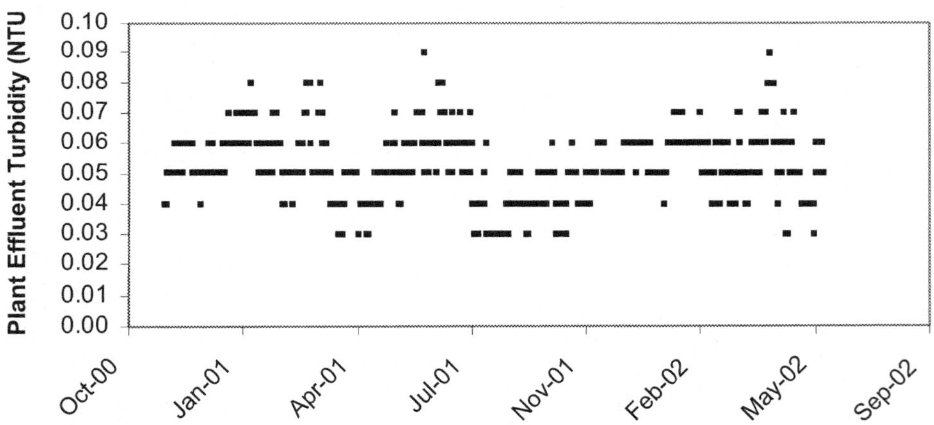

Figure 7.3. Average plant effluent daily turbidity values

Table 7.5
Individual filter turbidity (NTU) data [a]

Filter #	N[a]	Mean	STD	Min	25%ile	50%ile	75%ile	90%ile	95%ile	Max
Filter 1	574	0.034	0.01	0.01	0.03	0.03	0.04	0.05	0.05	0.07
Filter 2	567	0.038	0.01	0.02	0.03	0.04	0.05	0.05	0.06	0.13
Filter 3	542	0.042	0.02	0.02	0.03	0.04	0.05	0.06	0.07	0.16
Filter 4	557	0.041	0.02	0.02	0.03	0.04	0.05	0.06	0.06	0.42
Filter 5	537	0.044	0.02	0.01	0.03	0.04	0.05	0.06	0.07	0.38
Filter 6	554	0.044	0.01	0.01	0.04	0.04	0.05	0.06	0.06	0.10
Filter 7	543	0.036	0.01	0.01	0.03	0.03	0.04	0.05	0.06	0.11
Filter 8	568	0.047	0.01	0.02	0.04	0.05	0.05	0.06	0.07	0.14
Filter 9	540	0.040	0.01	0.02	0.03	0.04	0.05	0.06	0.06	0.16
Filter 10	556	0.040	0.01	0.02	0.03	0.04	0.05	0.06	0.06	0.16
Filter 11	511	0.051	0.01	0.02	0.04	0.05	0.06	0.07	0.08	0.10
Filter 12	538	0.039	0.01	0.01	0.03	0.04	0.04	0.05	0.06	0.10
Filter 13	523	0.048	0.01	0.02	0.04	0.05	0.05	0.07	0.07	0.14
Filter 14	553	0.047	0.02	0.02	0.04	0.04	0.05	0.07	0.07	0.15
Filter 15	569	0.045	0.01	0.02	0.04	0.04	0.05	0.06	0.07	0.11
Filter 16	576	0.050	0.01	0.00	0.04	0.05	0.06	0.07	0.07	0.11
Filter 17	570	0.051	0.01	0.02	0.04	0.05	0.06	0.07	0.08	0.12
Filter 18	545	0.051	0.02	0.00	0.04	0.04	0.05	0.06	0.07	0.40
Filter 19	539	0.055	0.01	0.03	0.05	0.06	0.06	0.07	0.08	0.11
Filter 20	529	0.037	0.02	0.01	0.03	0.04	0.04	0.06	0.06	0.09

[a] Based on average daily values.

Table 7.6.
Summary of individual filter turbidity values greater than 0.15 NTU

Date	Filter	Turbidity [a]
11/14/2000	18	0.40
11/20/2000	9	0.16
11/20/2000	10	0.16
11/27/2000	5	0.38
12/04/2000	4	0.42
12/20/2000	3	0.16

[a] Average daily values

The mean filter flow rate ranged from 0.87 to 0.94 MGD (minimum 0.49, maximum 1.59 MGD) (Table 7.7). The mean filter headloss for the 20 filters ranged from 1.02 to 1.54 ft (minimum range 0.0 to 0.50 ft, maximum range 3.05 to 4.08 ft) (Table 7.8). The average filter run time was 41.2 hours (range 16.7 to 48.8 hours) and the average number of filters washed per day was 11.1 (range 2 to 31 filters).

Particle count data (total counts >3μm) was collected daily from the plant effluent. The average plant effluent particle count was 78.41 particles/mL (range 3.4 to 1,510 particles/mL). The average difference between paired settled water values (after the Superpulsator) and the plant effluent was 1.5 logs. Plant effluent particle counts tended to be higher in the summer months probably due bacterial growth within the GAC filters (Figure 7.4). The settled particle count data showed no discernable seasonal trend.

The level of disinfection practiced at the plant is expressed as a ratio of the disinfectant concentration multiplied by the contact time (CxT) obtained, divided by the CxT required to achieve a 0.5 log inactivation credit for *Giardia* as required by the Surface Water Treatment Rule (USEPA 1989a). The average level of disinfection observed had a CxT ratio of 6.96 (range 1.01, to 29.4). The CxT ratio jumped in September 2001 when the new 2.47 MG clearwell was placed into service (Figure 7.5). The new clearwell is five times larger than the old clearwell (0.5 MG). A number of chemicals were used to treat the water at the East River Treatment Plant. These chemicals are summarized (by daily dose in pounds per million gallons) in Table 7.9. Caustic soda was used to maintain the pH at 7.4 (range 7.3 to 7.6), and the average alkalinity was 138 mg/L as $CaCO_3$ (range 85 to 164 mg/L as $CaCO_3$) and the average hardness was 186 mg/L as $CaCO_3$ (range 129 to 212 mg/L as $CaCO_3$) (Table 7.1). Zinc orthophosphate was added for corrosion control and the average phosphate concentration was 1.08 mg/L (range 0.86 to 1.26 mg/L). Fluoride was added in accordance with regulation and was maintained between 0.77 and 1.23 mg/L (average 1.00 mg/L). Total chlorine residuals in the treated plant effluent averaged 2.54 mg/L (range 2.14 to 2.94 mg/L). The average plant production was 17.5 MGD (range 13.0 to 27.0 MGD) (Table 7.1).

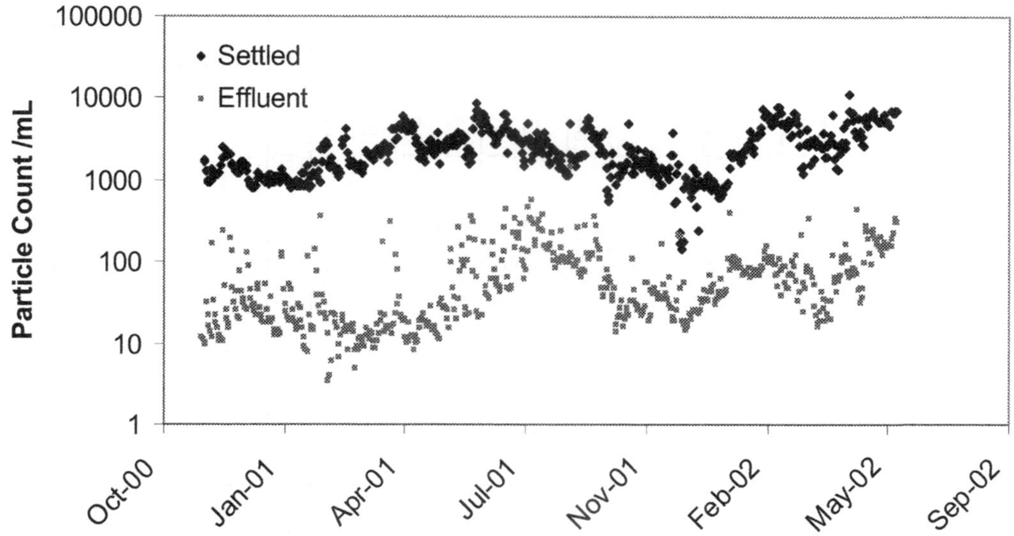

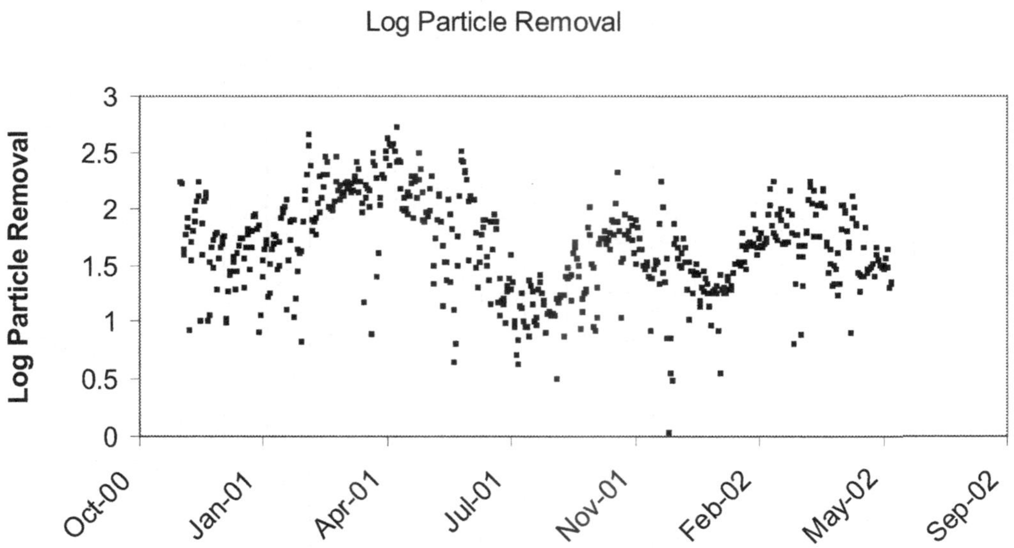

Figure 7.4. Comparison of settled water and plant effluent particle counts (top) and calculation of the log removal between settled water and plant effluent data (bottom).

Table 7.7.
Filter flow (MGD) rates for the East River Treatment Plant, Davenport, IA

Filter #	N[a]	Mean	STD	Min	25%ile	50%ile	75%ile	90%ile	95%ile	Max
Filter 1	574	0.88	0.15	0.51	0.80	0.86	0.93	1.07	1.20	1.42
Filter 2	567	0.87	0.15	0.50	0.80	0.86	0.93	1.06	1.17	1.55
Filter 3	544	0.87	0.16	0.51	0.80	0.85	0.92	1.09	1.21	1.55
Filter 4	557	0.87	0.15	0.52	0.79	0.86	0.92	1.04	1.17	1.47
Filter 5	535	0.87	0.16	0.50	0.80	0.86	0.93	1.09	1.22	1.44
Filter 6	565	0.87	0.15	0.47	0.80	0.86	0.93	1.06	1.19	1.48
Filter 7	542	0.86	0.14	0.54	0.80	0.86	0.92	1.01	1.15	1.38
Filter 8	568	0.87	0.15	0.50	0.80	0.86	0.925	1.04	1.20	1.48
Filter 9	540	0.87	0.14	0.51	0.81	0.86	0.92	1.04	1.19	1.42
Filter 10	557	0.87	0.15	0.52	0.80	0.86	0.92	1.05	1.18	1.56
Filter 11	511	0.92	0.15	0.49	0.83	0.89	0.99	1.15	1.24	1.56
Filter 12	534	0.93	0.15	0.64	0.84	0.90	0.98	1.18	1.25	1.58
Filter 13	518	0.94	0.16	0.54	0.83	0.89	1.00	1.19	1.29	1.58
Filter 14	551	0.93	0.16	0.61	0.83	0.89	0.99	1.18	1.27	1.57
Filter 15	569	0.94	0.16	0.66	0.84	0.89	0.99	1.18	1.29	1.57
Filter 16	576	0.93	0.16	0.00	0.84	0.89	0.99	1.18	1.26	1.59
Filter 17	570	0.94	0.16	0.65	0.83	0.89	0.98	1.175	1.28	1.58
Filter 18	545	0.92	0.17	0.00	0.83	0.89	0.98	1.14	1.24	1.56
Filter 19	546	0.94	0.16	0.63	0.83	0.90	0.99	1.18	1.29	1.57
Filter 20	525	0.93	0.15	0.68	0.83	0.89	0.99	1.17	1.28	1.57

Table 7.8
Filter headloss (ft) for the East River Treatment Plant, Davenport, IA

Filter #	N[a]	Mean	STD	Min	25%ile	50%ile	75%ile	90%ile	95%ile	Max
Filter 1	574	1.02	0.48	0.14	0.68	0.93	1.25	1.68	2.02	3.53
Filter 2	567	1.18	0.57	0.40	0.72	1.04	1.52	2.07	2.36	3.05
Filter 3	544	1.05	0.48	0.16	0.71	0.92	1.28	1.67	2.07	3.09
Filter 4	557	1.29	0.61	0.38	0.82	1.09	1.64	2.15	2.53	3.48
Filter 5	536	1.14	0.43	0.20	0.83	1.03	1.36	1.73	1.96	3.63
Filter 6	565	1.27	0.58	0.36	0.85	1.11	1.58	2.14	2.40	3.43
Filter 7	543	1.19	0.45	0.48	0.85	1.11	1.42	1.86	2.06	3.13
Filter 8	568	1.46	0.56	0.52	1.02	1.35	1.79	2.28	2.55	3.63
Filter 9	540	1.11	0.64	0.10	0.63	0.93	1.54	1.99	2.38	3.25
Filter 10	557	1.49	0.75	0.27	0.87	1.37	2.11	2.55	2.75	3.53
Filter 11	510	1.26	0.74	0.01	0.69	1.08	1.76	2.42	2.70	3.30
Filter 12	534	1.40	0.75	0.14	0.72	1.33	1.99	2.44	2.68	3.41
Filter 13	520	1.33	0.82	0.04	0.62	1.21	1.99	2.46	2.78	3.66
Filter 14	552	1.40	0.74	0.05	0.77	1.25	1.98	2.43	2.73	3.32
Filter 15	569	1.39	0.75	0.09	0.79	1.30	2.03	2.44	2.63	3.68
Filter 16	576	1.35	0.76	0.00	0.71	1.29	1.90	2.44	2.74	3.66
Filter 17	570	1.45	0.71	0.27	0.83	1.37	2.05	2.43	2.62	3.80
Filter 18	545	1.34	0.73	0.00	0.72	1.25	1.86	2.37	2.61	3.73
Filter 19	546	1.47	0.72	0.17	0.84	1.38	2.02	2.46	2.67	4.08
Filter 20	523	1.54	0.67	0.04	1.03	1.43	1.98	2.48	2.80	3.36

Table 7.9
Daily chemical dosages (LBS/MGD) for the East River Treatment Plant, Davenport, IA

Chemical Type	N[a]	Mean	STD	Min	25%ile	50%ile	75%ile	90%ile	95%ile	Max
Liquid fluoride	577	31.1	4.2	2.3	28.9	31.2	33.7	35.8	37.4	41.6
Dry fluoride	577	5.9	0.8	0.4	5.5	5.9	6.4	6.8	7.1	7.9
Gas liquid chlorine	577	52.7	13.7	27.2	40.6	50.9	62.6	72.3	77.8	86.4
Wet_PAS_A410	577	708	153	431	572	693	819	919	970	1260
Gas liquid ammonia	577	4.9	0.9	3.2	4.3	4.8	5.4	6.2	6.7	7.5
Dry_KMnO4	576	8.1	2.3	0.0	6.1	8.1	9.2	12.0	12.5	15.8
Salt Dry	364	7.7	3.5	0.0	5.8	7.3	9.9	12.3	12.9	19.3
Liquid zinc orthophosphate	577	18.2	2.4	12.5	16.8	17.9	19.2	21.1	23.5	28.0
Liquid solids	574	3.4	0.8	0.0	2.8	3.3	3.9	4.5	4.8	5.5
Liquid filter aid polymer	85	0.1	0.1	0.0	0.0	0.0	0.2	0.2	0.2	0.3
Wet caustic soda	459	140	76.0	4.2	73.9	123	215	250	264	288
Dry caustic soda	459	70.3	38.0	2.1	37.0	61.9	107	125	132	144
Cationic polymer	577	13.0	7.3	0.0	9.8	12.8	17.4	21.4	24.6	36.7

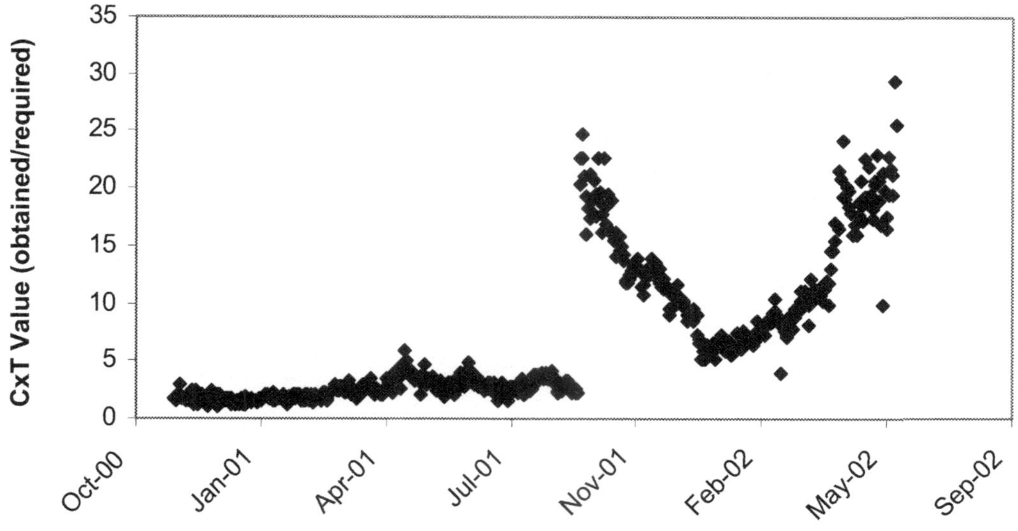

Figure 7.5. Calculated level of disinfection credit.
CxT values increased in September 2001 when the new clearwell was placed into service.

TREATMENT PERFORMANCE DURING THE 2001 FLOOD

Despite source water flood conditions between April 14 and May 23, 2001 (see Chapter 6), finished water quality remained excellent. The individual filter turbidities and combined filter effluent turbidities were consistently less than 0.1 NTU during this period. Filtered water particle count data were also low, averaging 18 particles/mL (range 11 to 38 particles/mL). The free chlorine CxT values were high, averaging 3.5 times higher than the required 0.5 log *Giardia* CxT. The plant effluent was consistently free of coliforms, infectious *Cryptosporidium* and viruses. Therefore, by all conventional methods of water quality monitoring, the finished drinking water was of very high quality.

CHAPTER 8
DISTRIBUTION SYSTEM RESULTS

The Iowa-American Water Company Davenport district distribution system includes more than 541 miles of mains, 5,053 hydrants, and varies in elevation from 565 feet to 862 feet above mean sea level. In the Main Service pressure zone, there is one storage facility, the 3.18 MG Ripley Street Reservoir. There are five elevated storage tanks with combined capacity of 4.1 MG in the High Service zone. There are seven booster pumping stations in the distribution system that transfer water between areas or raise supplies to more distant higher elevation areas.

Monitoring of the distribution system was based on the existing State of Iowa-approved coliform monitoring plan that collects approximately 120 samples per month. The distribution system is divided into four regions (labeled A through D) that are tested Monday through Friday. Normally 6 to 8 samples are collected daily, and analyzed for total coliforms, HPC bacteria and total chlorine. All locations are in public buildings, and are typically restrooms or utility rooms of gas stations, restaurants, schools, hotels, or businesses (Figure 8.1). A more complete suite of tests were performed from hospitals centrally located within the system. This suite included temperature, pH, phosphate, free and total chlorine, male-specific and somatic coliphage as well as total coliform and HPC analyses. A summary of distribution system physiochemical and biological parameters is shown in Table 8.1.

Figure 8.1. Representative distribution system sampling locations

ANALYSIS OF MICROBIAL INDICATORS

A total of 2,471 coliform samples were collected during the study, and only two samples tested positive (0.08%) for total coliform bacteria. These positive samples were collected on 10/15/01 and 12/11/01, and had counts of 1 and 50 cfu/100 mL, respectively (Table 8.2). Residual total chlorine in these samples was 1.9 and 1.6 mg/L, respectively. The samples were negative for *Escherichia coli* in EC-MUG broth. Repeat samples of sites within 5 service connections upstream and downstream of the positive site were also negative for total coliform bacteria and *E. coli*. Based on these data it is assumed that the total coliform results were due to localized contamination of the sampling faucet and were not representative of water quality in the distribution system. The overall average concentration of distribution system total coliform bacteria during the study was 0.02 cfu/100 mL.

A total of 2,430 samples were examined for HPC bacteria (Table 8.1) and the average HPC was 230 cfu/mL (range 0 to 2,300 cfu/mL). The average distribution system total chlorine residual was 2.08 mg/L (range 0.59 to 3.60 mg/L), while the average free chlorine was 0.06 mg/L. The DPD test for free chlorine produces a slight false reaction when high levels of chloramines are present (APHA 1998). Free chlorine was not likely detectable in these samples, and Iowa-American did not convert to free chlorine during the study for flushing or nitrification control. There was no correlation between the concentration of HPC bacteria and total chlorine residuals (Figure 8.2), and no observable seasonal pattern in HPC occurrence (Figure 8.3).

Male-specific coliphage were detected in 22 of 393 (5.6%) samples examined (Table 8.1), with an average concentration of 0.1 pfu/100 mL (range 0 to 5 pfu/100 mL). Table 8.3 provides a summary of all distribution system male-specific coliphage positive samples. Fifteen distribution system samples were excluded from the analysis because of a contaminated control (see Chapter 5). The positive samples occurred during the winter months, in December 2000 through March 2001 and again November 2001 through February 2002. The only positive somatic coliphage samples occurred during the period when the control was contaminated; providing strong justification to invalidate these results.

Phage serotyping is used for tracking sources of contamination. An experiment was conducted to determine whether the male-specific phages detected in distribution system samples contained a DNA or RNA genome. To do this, isolates were exposed to enzymes that degrade RNA. Those phage resistant to the enzymes contained DNA, and isolates sensitive to the enzymes contained RNA. All distribution system coliphage contained RNA (Table 8.4). RNA containing male-specific coliphage are composed of four serogroups (I-IV). Strains isolated from human feces are typically groups II and III, while groups I and IV are usually found in animal feces. Serotyping was performed by Dr. Jill Stewart of the National Oceanic & Atmospheric Administration (NOAA). All isolates were reported to be serotype 1 (Table 8.4). Although the data suggest that the phage did not originate from a human fecal sample, it is unlikely that the organisms could have grown in distribution system biofilms.

Table 8.1

Physiochemical and biological water quality parameters for the Iowa-American Water Company distribution system

Parameter	N[a]	N < DL[b]	Mean	STD	Min	25%ile	50%ile	75%ile	90%ile	95%ile	Max
pH (pH units)	358		7.4	0.1	7.2	7.3	7.4	7.5	7.6	7.6	7.9
Turbidity (ntu)	394		0.14	0.03	0.08	0.12	0.14	0.16	0.18	0.19	0.21
Temperature (°C)	358		13.63	8.57	3.00	6.00	12.00	20.00	28.00	29.00	31.00
Phosphate (mg/L)	394		1.20	0.12	0.83	1.12	1.20	1.28	1.35	1.38	1.56
Free chlorine (mg/L)	358		0.06	0.05	0.02	0.03	0.05	0.05	0.10	0.15	0.30
Total chlorine (mg/L)	2471		2.08	0.34	0.59	1.90	2.10	2.30	2.50	2.60	3.60
Total coliform (cfu/100 mL)	2471	2469	0.02	1.0	0	0	0	0	0	0	50
Male specific coliphage (pfu/100 mL)[c]	393	371	0.1	0.5	0	0	0	0	0	1	5
Somatic coliphage (pfu/100 mL)[d]	399	399	0	0	0	0	0	0	0	0	0
HPC (cfu/mL)	2430	426	231.9	432.1	0	1	40	190	845	1430	2260
Fluoride (mg/L)[e]	377		0.97	0.12	0.00	0.91	1.00	1.03	1.08	1.11	1.24
Phosphate (mg/L)[e]	394		1.20	0.11	0.83	1.12	1.20	1.28	1.35	1.38	1.56
Total Chlorine (mg/L)[e]	394		2.10	0.24	1.22	1.95	2.10	2.23	2.37	2.50	2.92

[a] Number of samples [b] Number of samples below the detection limit i.e. not detected. [c] 15 Male Specific analyses were contaminated and they were excluded from the data analysis; 6 samples were lost and not processed. [d] 15 somatic analyses were contaminated and excluded from the data analysis. [e] From representative distribution system sites (hospitals) that were sampled daily.

Table 8.2
Distribution system total coliform results

Sample ID	Sample date	CFU/100mL
28801A011	10/15/01	1
34501A011	12/11/01	50

Table 8.3
Distribution system male-specific coliphage results

Sample ID	Sample	PFU/100mL	DS location
34700A013	12/12/2000	1	A
34700C009	12/12/2000	1	C
00201B029	01/02/2001	5	B
00201D004	01/02/2001	3	D
00201B001	01/02/2001	2	B
00901B020	01/09/2001	2	B
02301B029	01/23/2001	1	B
03001C004	01/30/2001	1	C
03701D011	02/06/2001	4	D
03701B012	02/06/2001	1	B
04401D016	02/13/2001	1	D
05801A011	02/27/2001	1	A
06501A013	03/06/2001	4	A
06501A011	03/06/2001	2	A
06501B029	03/06/2001	2	B
07201D016	03/13/2001	1	D
07901D005	03/20/2001	2	D
31001A011	11/06/2001	2	A
31701C009	11/13/2001	2	C
31701B024	11/13/2001	1	B
31701D009	11/13/2001	1	D
05002D009	02/19/2002	1	D

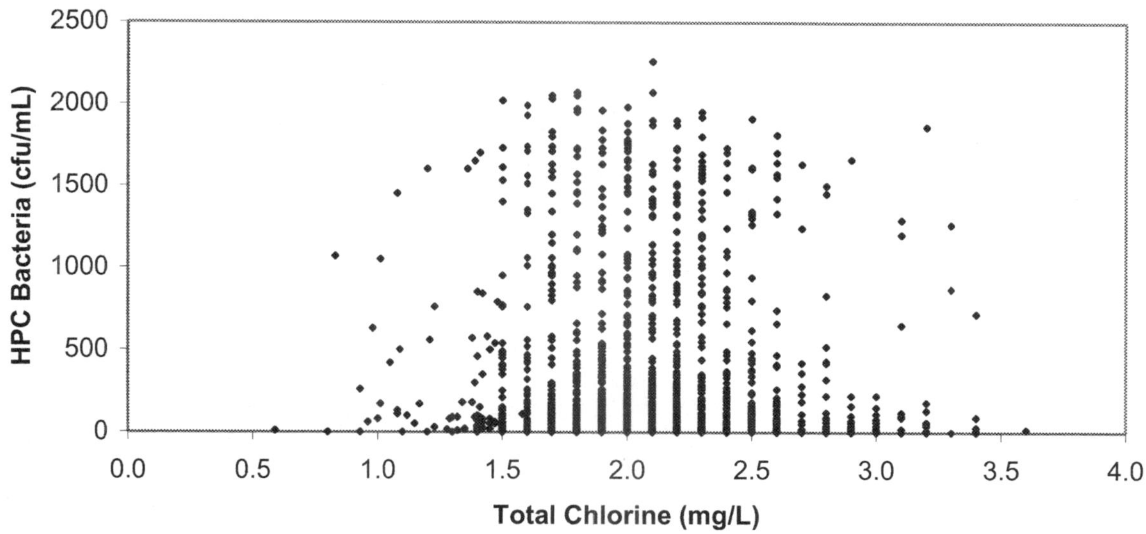

Figure 8.2. Relationship between distribution system HPC bacteria and total chlorine (chloramine) residual.

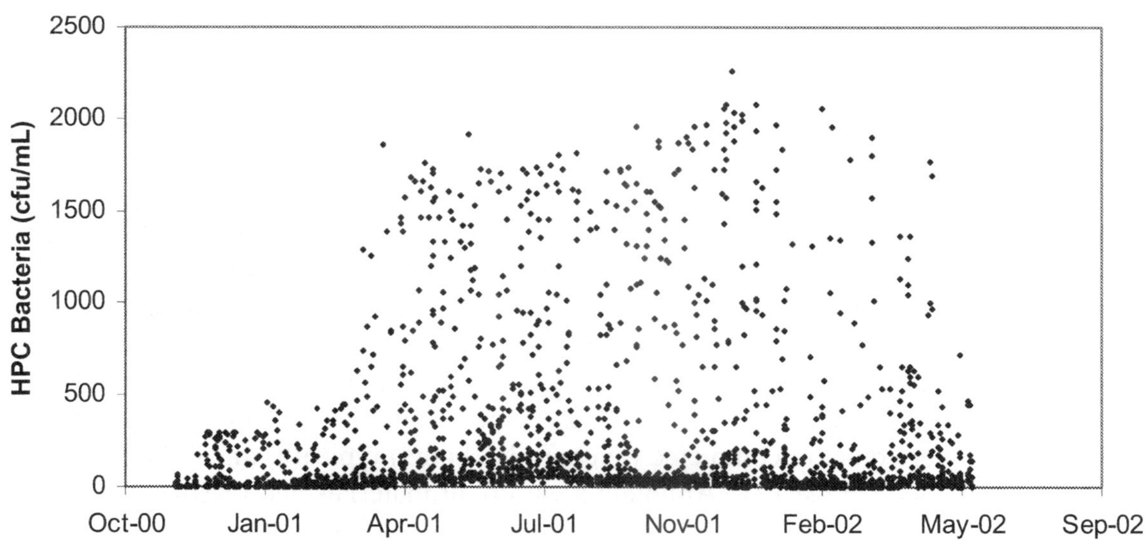

Figure 8.3. Seasonal occurrence of distribution system HPC bacteria.

Table 8.4
Serotyping of distribution system male-specific coliphage

Sample ID	RNase positive spot plate test	RNase negative spot plate test	RNA/DNA phage	Serotype
05002D009	-	+	RNA	1
31701BO24	-	+	RNA	1
31701C009-A*	-	+	RNA	1
31701C009-B*	-	+	RNA	1
31701D009	-	+	RNA	1
31001A011-A*	-	+	RNA	1
31001A011-B*	-	+	RNA	1
31001B020	-	+	RNA	1
MS-2 stock (positive control)	-	+	RNA	ND
Positive control	-	+	RNA	ND
Negative control	-	-	-	ND
DNA Phage (PhiX-174)	+	+	DNA	ND

* A, and B indicates that more than one plaque was purified from a single agar layer plate. ND, not done.

ANALYSIS OF PHYSIOCHEMICAL PARAMETERS

A total of 394 samples were collected for physiochemical parameters including phosphate, turbidity, pH (358), temperature (358), and free chlorine (358). The average distribution system pH was 7.4 (range 7.2 to 7.9) and the average temperature was 13.6°C (range 3.0 to 31°C). The average turbidity of distribution system samples was 0.14 ntu (range 0.08 to 0.21 NTU), and phosphate levels (indicative of the zinc orthophosphate corrosion inhibitor) averaged 1.20 mg/L (range 0.83 to 1.56 mg/L). Statistical comparison of the distribution system results to the plant effluent and household values is provided in Chapter 10.

SUMMARY OF MAIN BREAKS AND DISTRIBUTION SYSTEM REPAIRS

Distribution system work orders processed by Iowa American generate a log of all distribution system (observed) leaks and repairs. Analysis of historical records indicates that most leaks occurred during the winter months (November to February) when freezing temperatures result in frost heave (Figure 8.4). The winter of 2000 was particularly cold with a record number of main breaks in November and December 2000 (27 and 36, respectively) when water temperatures plummeted from 16.7°C to 1°C. In addition, July and August 2001 had an unusually high number of repairs (11 and 12, respectively) for the typically quiet summer months. Repairs were quickly completed (average 1.5 hours) typically impacting only a limited number of customers (average 12). Although a variety of materials and reasons were associated with main breaks, most repairs were performed on 8-inch cast iron pipes with a circular type of break associated with pipe movement.

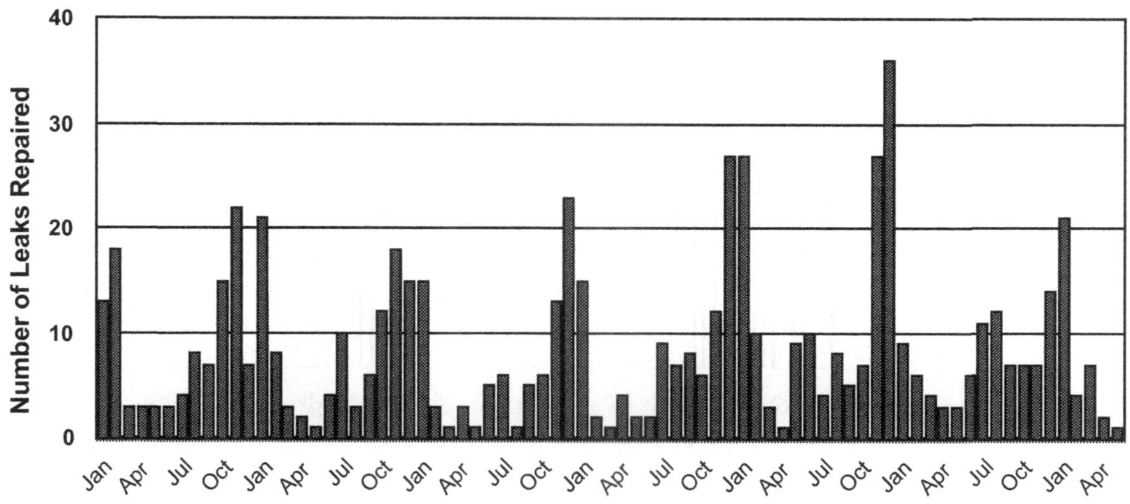

Figure 8.4. Seasonal distribution of leak repairs

Because coliform bacteria were detected on only two occasions (Table 8.2) there was no evidence that main breaks resulted in coliform contamination of the distribution system. Most repairs were conducted under pressure, thus limiting the potential for contamination to the distribution system. When the pipelines are depressurized for repair, the pipe sections are isolated and disinfected with free chlorine according to AWWA protocols (AWWA C651, 1992) prior to returning the area to service. The occurrence of male-specific coliphage in December 2000 through March 2001 (Figure 8.5) may have been associated with the unusually large number of main breaks in late 2000, although the two events are not closely related. The poor effectiveness of residual chloramines for virus inactivation may have contributed to the survival of the viruses in the distribution system or in biofilms. Because the coliphage serotype was not associated with human fecal pollution, the public health significance of the results is not clear.

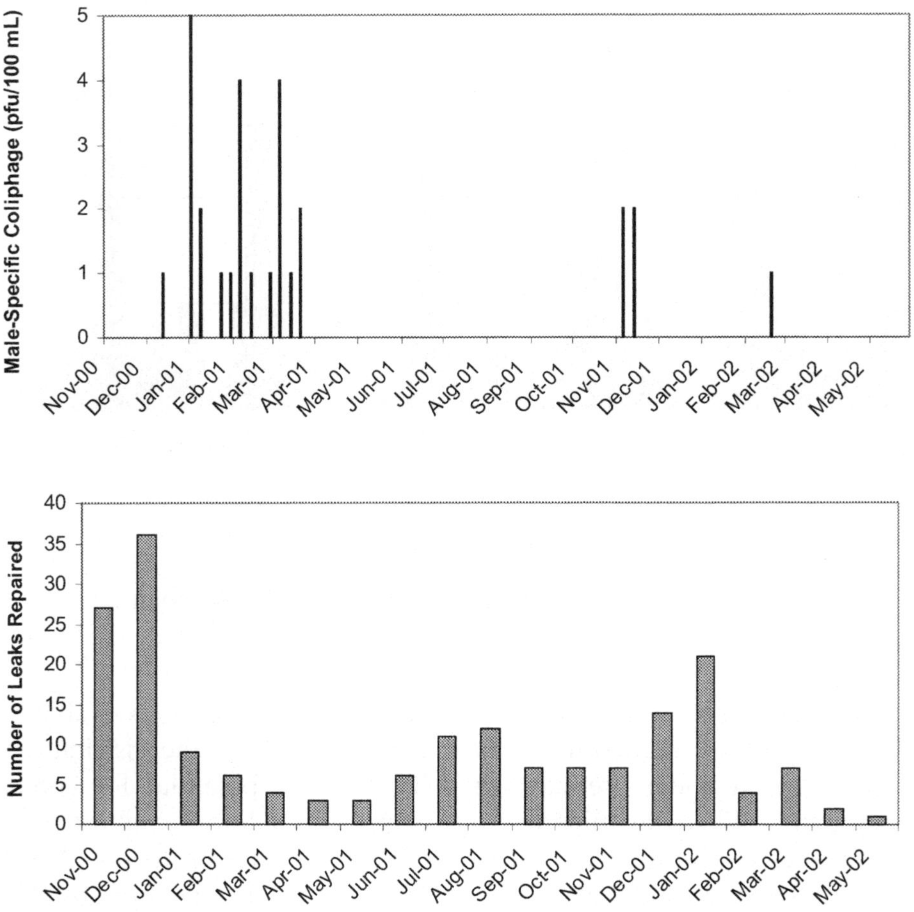

Figure 8.5. Association between the occurrence of distribution system male-specific coliphage (top) and pipeline repairs (bottom).

CHAPTER 9
OCCURRENCE AND MODELING OF PRESSURE TRANSIENTS

Detailed monitoring of distribution system pressure and hydraulic modeling of the pipe network was performed to supplement the microbial monitoring outlined in Chapter 8. Maintenance of a positive distribution system pressure is an important control measure to prevent the introduction of contaminants external to the pipe network. Continuous monitoring of distribution system pressures ensured the maintenance of this barrier in between discrete microbiological monitoring events. Application of hydraulic modeling was used to extend this evaluation to all parts of the distribution system. Modeling was also used to evaluate various operating scenarios and examine potential control strategies.

PRESSURE MONITORING

The Davenport distribution system is comprised of seven pressure zones and an electronic pressure monitor was placed in each zone (Figure 9.1). Pressure Zone #1 and Zone #3 are supplied from the treatment plant via separate sets of Main and High Service pumps at average discharge pressures of 73 and 112 pounds per square inch (psi), respectively. Zone #1 encompasses the lower elevation area along the Mississippi River. Zone #3 includes higher elevations along the river bluff, and is the larger and faster-growing part of the system. A booster pump in Zone #1 transfers water to Zone #4, and another pumps to an elevated tank. The Ripley booster station, next to the elevated tank, pumps water from the tank into Zone #3 (monitoring site #3 is located close to the tank and booster station). Four other booster stations in Zone #3 are used to supply water to the remaining four pressure zones. In total, there are five discharge stations (points where water is transferred) in Zone #3 and two in Zone #1, and four storage facilities in Zone #3, one in Zone #1, and one in Zone #5.

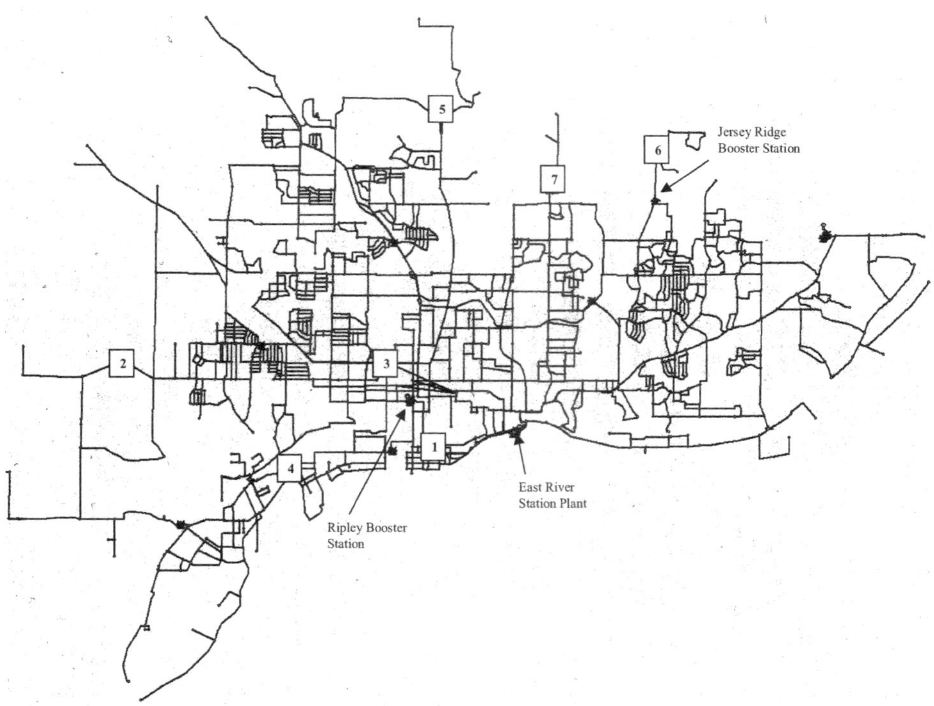

Figure 9.1. Schematic of the Davenport distribution system. Numbers indicate monitoring locations and pressure zones.

Figure 9.2. Installation of a pressure monitor on a fire line.

The electronic pressure monitor used was a high-speed single-channel pressure transient data logger (Model RDL 1071L/3 Pressure Transient Logger; Radcom Technologies, Inc., Woburn, MA). The units were placed at in-door locations where they could be connected to a fire line so to collect distribution system pressures without exposure to freezing conditions (Figure 9.2). The Radcom monitors were installed and set up according to manufacturer's instructions. The monitors were zeroed to atmospheric pressure upon each installation. To minimize the amount of data collected, the recorders were set to take one reading per second at a tolerance of ±4.0 psi. With this setting, the monitor recorded new pressure readings only if the values changed by more than ±4.0 psi. This enabled about a week's worth of data to be collected before the recorder memory became full and no further data would be recorded. The clock time used by the logger was periodically calibrated by use of a connection to a personal computer. The times were used to relate the pressure data to distribution system events. A summary of the pressure monitoring parameters is presented in Table 9.1, and Appendix G contains a summary of all the pressure data.

Table 9.1
Summary of pressure monitoring parameters

Site No.	Days of Available Data	Dates monitored	Description of Monitoring	Tolerance Setting (+/- psi)	Readings per Second	No. of Pressure events <20 psi	Observed pressure range
1	428	3/19/01 – 6/3/02	Zone #1	4.0	5	0	29 – 96
2	439	3/22/01 – 6/3/02	Zone #2	4.0	5	0	27 – 76
3	438	3/15/01 – 10/11/02	Zone #3	4.0	5	4	9.5 – 67
4	508	3/23/01 – 6/3/02	Zone #4	4.0	5	1	19.8 – 107
5	432	3/16/01 – 6/3/02	Zone #5	4.0	5	0	44 – 103
6	439	3/15/01 – 6/3/02	Zone #6	4.0	5	0	31 – 89
7	506	3/16/01 – 10/11/02	Zone #7	4.0	5	0	31 - 105

Although the project ended on June 1, 2002, additional pressure data was collected until October 11, 2002 in an attempt to capture negative pressure events. Based on a combined 3,190 days of pressure monitoring, totaling 2.756×10^8 data points, only five occasions were observed where distribution system pressures were less than 20 psi, and no negative pressures were observed.

A summary of the five occasions where distribution system pressure over the 18-month study period was less than 20 psi is shown in Table 9.2. In each case, the pressure dip (<20 psi) was observed in only one pressure zone, though pressure transients were simultaneously experienced in some, or all, of the other pressure zones. Pressure Zone #3 (the high service zone) experienced the most frequent surges as a result of routine pump shutdowns or power failures (Table 9.2). In addition, a pipe break in zone #4 caused pressures to drop below 20 psi. Other than these events, no pressures below 20 psi were detected at any of the monitoring sites in the Davenport distribution system for the over 18 months of the study period. The recorded pressures were above 27 psi during the monitoring period at Sites #1, #2, #5, #6, and #7 (Table 9.1), and the results for these sites are not further discussed.

Table 9.2
Observed pressures less than 20 psi

Event No.	Date	Site/Zone ID Number	Initial Pressure (psi)	Minimum Pressure (psi)	Pressure Differential (psi)	Time <20 psi (sec)	Suspected Cause of Surge	Figure No.
1	5/17/01	3	51.3	17.3	34.0	8	Unknown *	---
2	7/31/01	3	49.9	19.9	30.0	7	Unknown *	9.3
3	8/2/01	3	51.0	9.5	41.5	19	Power outage (lightning)	9.3
4	9/2/01	3	50.7	18.3	32.4	1-2	Unknown *	---
5	8/10/01	4	71.2	19.8	51.4	2,580	Hole bored in service line	9.4

* Assumed to be due to regular pump shutdown

Event 1. May 17, 2001

At 2:59 a.m. on May 17, 2001 a pressure event was experienced at Site #3 with the pressure dropping from 51.3 psi to 17.3 psi for 8 seconds. No explanation was found for the cause of the pressure event, but was assumed to be due to regular pump shutdown. The monitoring location is in the High Service pressure zone, and is approximately 5 blocks from the Ripley booster station (which provides water from an elevated storage tank supplied by the Main Service zone). At the same time, there were also smaller pressure drops at other sites, including Site #1, #4, and Site #7. Site #6 showed a very slight drop in pressure and two sites showed no change in pressure (Site #2 and #5).

Event 2. July 31, 2001

At 8:13 p.m. on July 31, 2001 a pressure event was experienced at Site #3 with the pressure dropping from 49.2 psi to 19.9 psi for 7 seconds. No explanation for the pressure drop was determined, but was assumed to be due to regular pump shutdown. The event was not unusual as several other transients were in the range of 25 to 30 psi, all apparently caused by the same mechanism. At the same time, there was also a lesser pressure drop at Site #1, but sites #4, #6, and #7 showed only slight changes in pressure, and Site #2 showed almost no change in pressure. Flow to the pipe at Site #5 had been shut off and there are no pressure results available for that site.

Event 3. August 2, 2001

In other on-going studies of pressure variations (AwwaRF 2686), most of the significant downward transients were caused by pumps turning off. One such event is shown in Figure 9.3 as Event #3, where a power outage occurred for six minutes at the treatment plant, shutting the high service pumps down. The pump at the Ripley booster station near Site #3 apparently also lost power. The pressure dropped from 51.0 psi to 9.5 psi for 19 seconds (Figure 9.3). Monitors at the other locations recorded some small pressure variations, but not substantially different than the normal variations recorded on a routine basis. Analysis of pressure recordings during other power outages did not produce low-pressure events.

Event 4. September 2, 2001

At 6:12 a.m. on September 2, 2001 a pressure event was experienced at Site #3 with the pressure dropping from 50.7 psi to 18.3 psi for 1-2 seconds. The cause of this sudden decrease is not known, but was assumed to be due to regular pump shutdown. Sites 1 and 4 showed smaller but noticeable pressure drops, although they were not larger than other pressure variations recorded during that period.

Event 5. August 10, 2001

The pressure event shown in Figure 9.4 apparently occurred when a contractor inadvertently drilled a hole through a 2-inch water service line in pressure Zone #4 on August 10, 2001. The pressure dropped from 71.2 psi to 19.8 psi and stayed at 19.8 (±4.0) psi for 43 minutes before slowly increasing to normal. This 2-inch line is about 20 feet long, and the hole was drilled at about the midpoint of it, approximately ten feet from a 16-inch main, which is between the curb and the sidewalk. Three 1-inch lines (corporations) are connected to the main within a 4-foot area, and these three lines tie into a Siamese connection going into a roadway box. The 2-inch line that had the hole comes out from this roadway box (and is the only pipe to do so). Site 1 showed a pressure drop at this time, although it was no larger than the normal pressure variation at this site. No discernible pressure changes occurred at Sites 2, 3, 6, or 7 (Site 5 was turned off).

DISTRIBUTION SYSTEM MODELING

Hydraulic Model Development

Data from the existing Iowa American Water Company, Davenport District Cybernet model was converted to the KYPipe2000 format by researchers at the University of Kentucky.

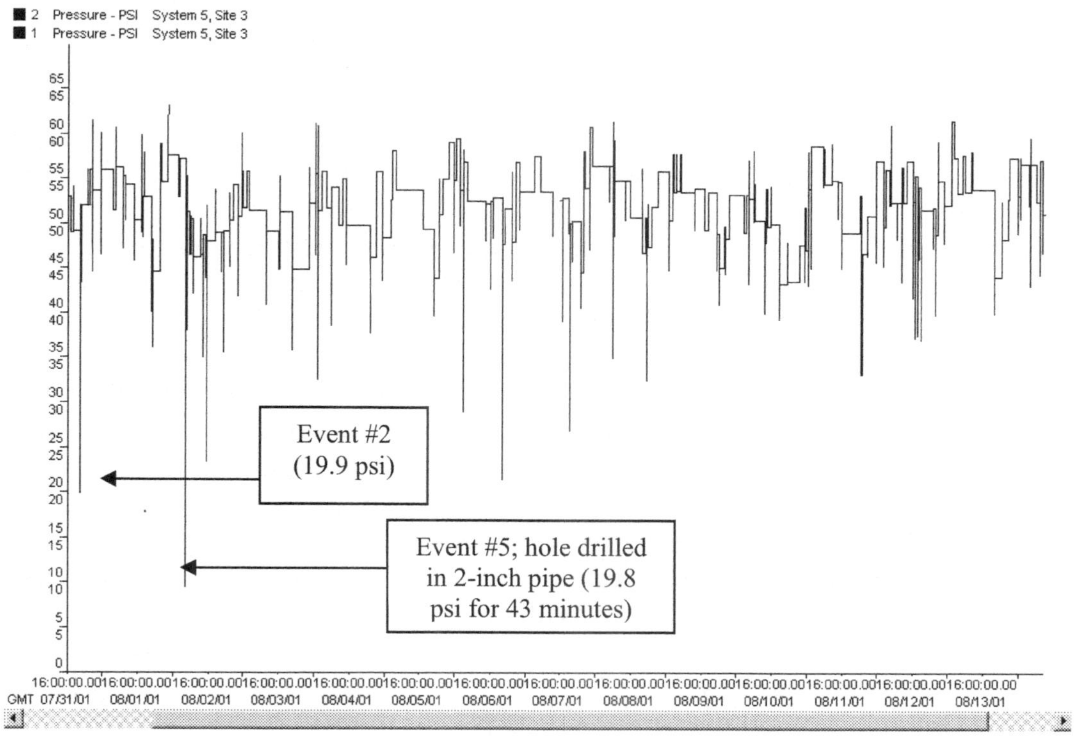

Figure 9.3. Pressure recording from zone 3 showing two events where pressures dipped to <20 psi.

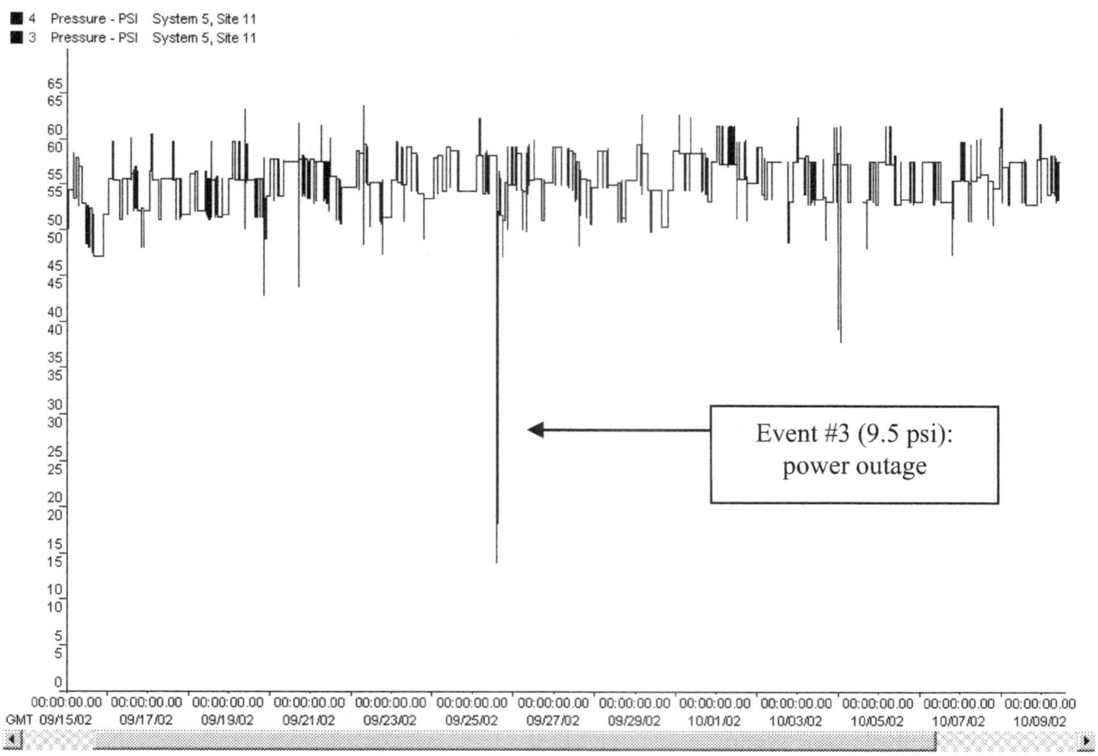

Figure 9.4. Low pressure apparently caused when a contractor drilled a hole through a 2-inch water service line.

The model is comprised of 1703 pipes, 1146 nodes, 12 supplies, 30 pumps, and includes 6, 8, 12, 16, 24 and >24 inch mains. Water demand data used 1999 results. The two models were aligned and compared, and the data for the models were in good agreement. It was concluded that the Pipe2000 model was virtually identical to the Cybernet model and the pressures agreed within ± 1 psi. A schematic of the Pipe2000 model is shown in Figure 9.5 and is color-coded to show pipe diameters. Elevations in the system are shown in Figure 9.6.

The Pipe2000 model was upgraded to incorporate any modifications or additions to the distribution system. These changes largely include addition of new pipe sections, the Jersey Ridge Booster Station, and the removal of some unused (or not installed) pipes. Fire flow data for 17 tests were provided to the University of Kentucky modelers for calibration of the model (Table 9.3, Figure 9.7). For 6 of these tests, the static pressure head obtained in the fire flow test corresponded well to the model residual pressures and were used to optimize the model. The other test results were considered not valid because some other condition was different – either the elevations were incorrect or some other boundary condition was not accounted for (e.g., pump status, tank level. etc.). Using the fire flow test results, a formal calibration was carried out using the Pipe2000 Optimized Calibration module. To accomplish this, five pipe groups were selected based on the values for Hazen Williams C values assigned for the original model.

Group 1	30-49
Group 2	50-69
Group 3	70-89
Group 4	90-109
Group 5	110-130

Calibration is affected by the demands and measured fire flows and the values used always have some uncertainty. An accuracy of 10% was assumed for these parameters. This allows the calibration to utilize this uncertainty when making adjustments. The calibration can be done in one of two ways:

1. A single optimum Hazen Williams Coefficient can be calculated for each group.
2. An optimum multiplication factor for the assigned roughness in each group can be determined.

The second approach provides slightly better results and is preferable because the trends due to the assigned Hazen Williams coefficients can be maintained within each group. For example, pipes assigned Hazen Williams coefficients differing by 10% were assigned new values that also differed by 10%. These results were used to adjust the Davenport model data. The data differences are limited to roughness adjustments for the five pipe groups. These adjustments (multiplication factors) were as follows:

Group 1	1.10
Group 2	1.20
Group 3	1.15
Group 4	.92
Group 5	.90

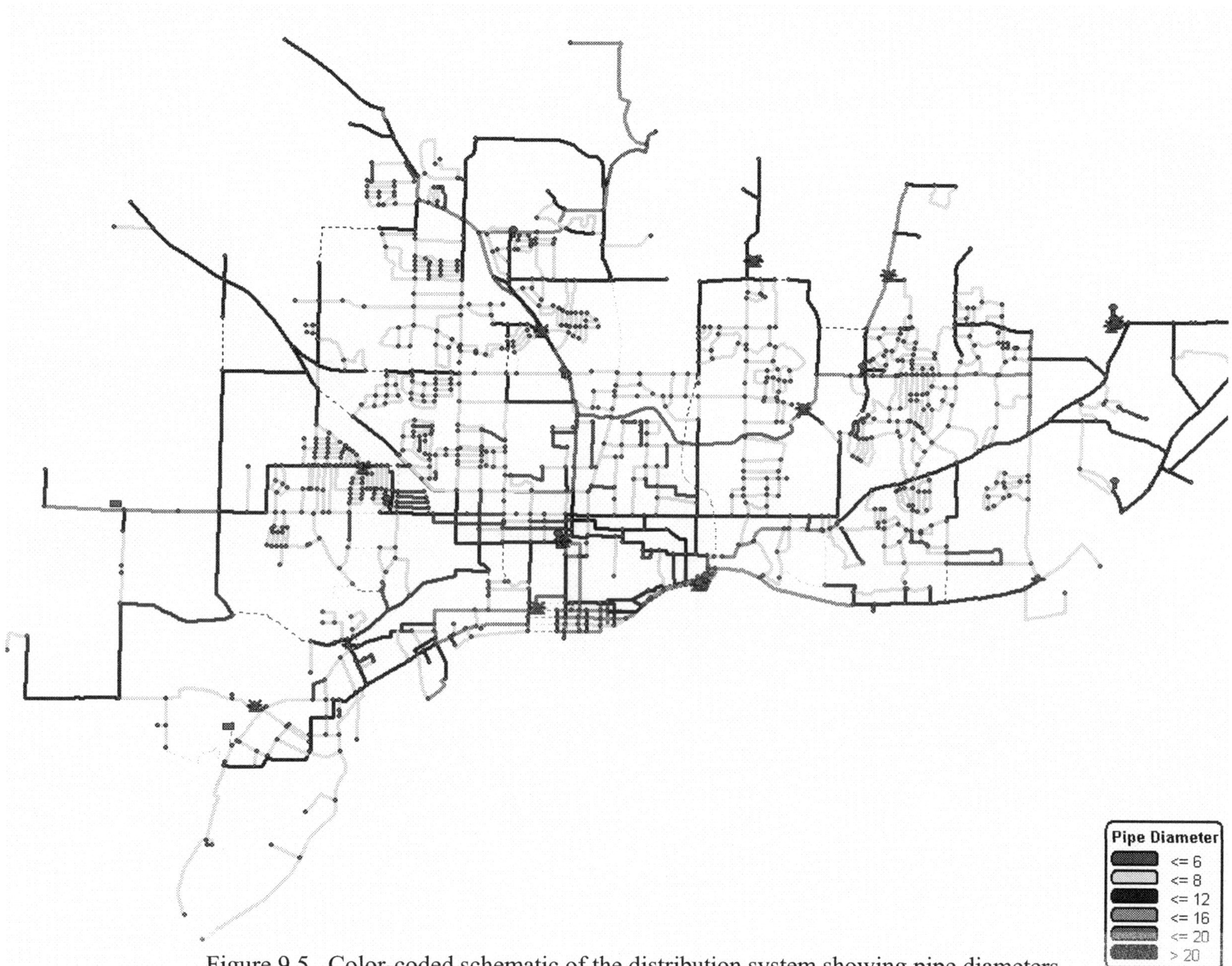

Figure 9.5. Color-coded schematic of the distribution system showing pipe diameters.

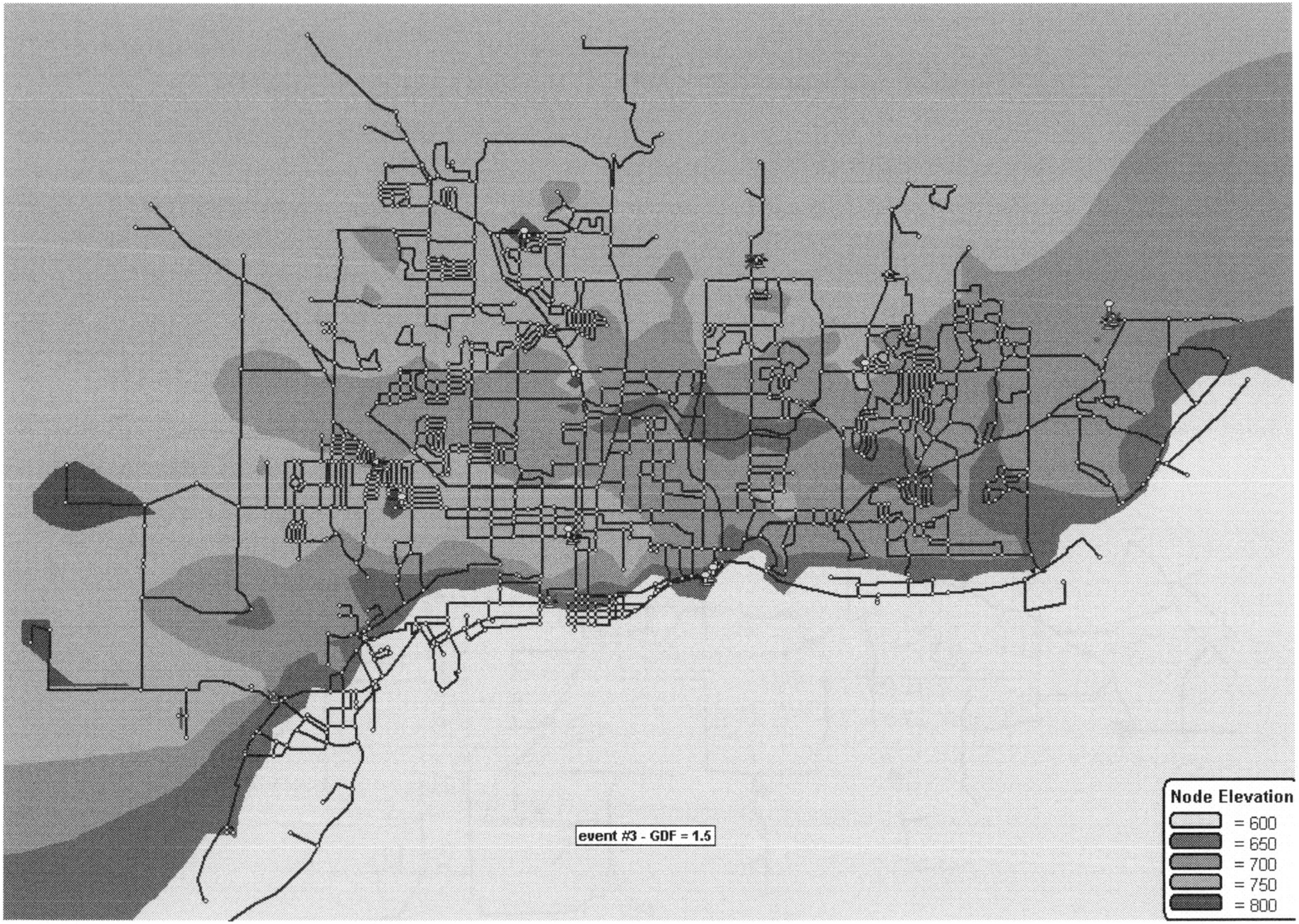

Figure 9.6. Elevations in the Davenport distribution system.

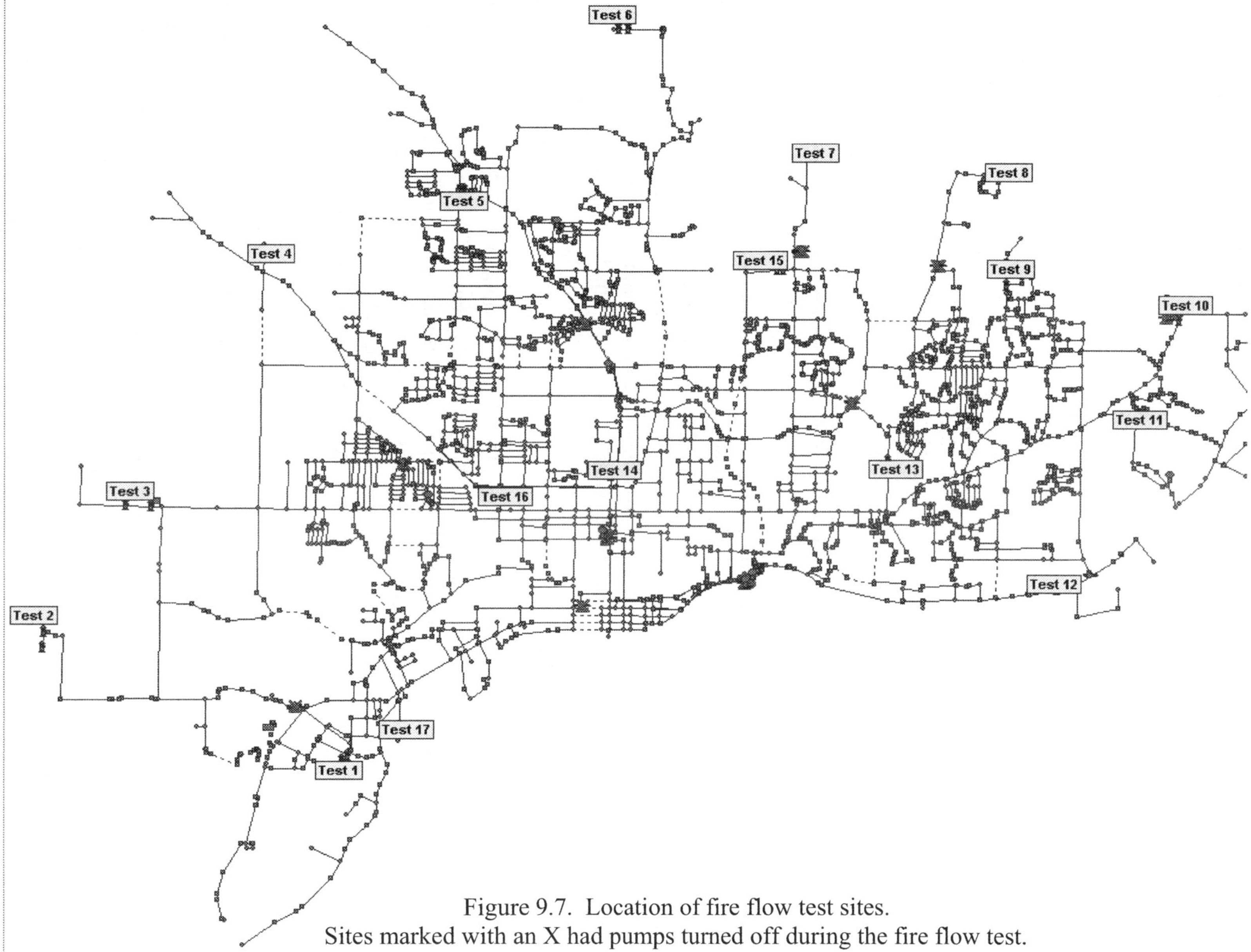

Figure 9.7. Location of fire flow test sites.
Sites marked with an X had pumps turned off during the fire flow test.

Table 9.3
Fire Flow Test Data for Davenport

Test	Elevation	Measured static pressure	Measured HGL (ft.)	Pipe2000 static pressure (psi)	Model HGL (ft.)
1	570	62	713.1	68	726.9
2	750	48	860.8	42	846.9
3	750	49	863.1	49	863.1
4	735	60	873.5	49	848.1
5*	735	52	855.0	52	855
6	730	50	845.4	56	859.2
7	730	75	903.1	43	829.2
8*	730	72	896.2	38	817.7
9*	700	46	806.2	50	815.4
10*	695	54	819.6	51	812.7
11*	682	59	818.2	57	813.6
12	585	60	723.5	71	748.9
13	665	85	861.2	70	826.6
14	685	58	818.9	63	830.4
15*	695	58	828.9	56	824.2
16	710	48	820.8	50	825.4
17	~560	41	654.6	71	723.9

* Indicates flow data used to calibrate the model

The results of a hydraulic calculation showed that the differences caused by the adjusted Hazen Williams values were very small and the pressures varied by 2 psi or less. The obvious conclusion is that the original model was reasonably well calibrated and the Hazen Williams Coefficient values were well chosen. A second conclusion is that the Davenport distribution system operates with relatively small head loss under baseline conditions and, for this situation, the sensitivity to Hazen Williams Coefficient values is relatively small.

Davenport Surge 2000 Model

The calibrated Pipe2000 model of the Davenport System was utilized to create the Surge2000 model. Prior to creating the Surge2000 model, the status of approximately 25 closed pipes was verified and all permanently closed pipes were eliminated from the model. Several pipes that had been designated as closed were actually opened. The pressures at selected nodes were compared before and after the adjustments involving closed pipes. Some slight increases in pressure were noted in the western side of the system due to a few pipes that were originally designated as closed but were actually opened.

A wave speed of 3000 feet/second was assigned to all pipes, which is a reasonable value for metal and reinforced pipes with a small amount of entrained air. To assure that the Surge2000 model was initially balanced and holding the initial steady state conditions, a ten second transient was carried out with no transient producing conditions specified. The results showed the maximum/minimum pressures obtained by the transient analysis for the ten second run and the maximums and minimums were all identical, verifying that the model was starting with balanced conditions. This is an important verification check for surge models.

Modeling Surge Events

A review of the five surge events revealed that for only Event 3 (08/02/2001) and Event 5 (08/10/2001) could the cause of the surge event be determined (Table 9.2). For Event 3, a power outage at the main pumping facility (East River Station) produced a down surge that was recorded at six of the seven monitoring stations. For Event 5, a leak (broken pipe) near Site 4 lowered the pressure to around 20 psi for an extended period.

A total of 8 different model scenarios were performed for the Event #3 power outage, depending on the number of pumps in service (it was not known exactly how many pumps were in service at the time of the lighting strike) and the system demand at the time (Table 9.4). The scenarios examine the combinations of 1 Main Service and 3 High Service pumps and a global demand factor (GDF) of 0.5, 1.0 and 1.5; and a higher operating rate with 2 Main Service and 4 High Service pumps and a GDF of 2.0. Because there were indications that the Ripley booster station may have also lost power, a modeling scenario was performed with this station either off or on. A fourth case study was performed where two surge tanks (closed 20-foot diameter tanks with 10 feet of water and 10 feet of air in the surge tank) were modeled downstream from the Main and High Service pumps stations.

Table 9.4
Summary of Event #3 Modeling Scenarios

Case Study	Description		
	Number of Pump Trips	Global Demand Factor	Surge Tank
1	1 Main Service, 3 High Service, Ripley booster	1.0	No
1a	1 Main Service, 3 High Service, Ripley booster	1.5	No
1b	1 Main Service, 3 High Service, Ripley booster	0.5	No
2	2 Main Service, 4 High Service, Ripley booster	2.0	No
3	1 Main Service, 3 High Service,	1.0	No
4	2 Main Service, 4 High Service, Ripley booster	1.0	Yes
5	All pumps	1.0	No
6	All pumps	1.0	Yes

For each of the modeling scenarios (Case studies 1-3), the calculated pressures matched reasonably well with the observed pressures monitored at the seven recording stations (Table 9.5). Figure 9.8 shows that the model output matched closely the observed low-pressure transient. In each case, constraining the model to the observed values ensured that the output results match the actual conditions. As expected, the range of pressure variation was slightly greater in Case 2 where the system demand was greater (GDF of 2.0). The output for the various case studies varied only slightly in the number of nodes that experienced negative pressures, and a representative diagram is shown in Figure 9.9. For all 5 scenarios (Case studies 1-3), three principle areas of the distribution system (circled) were projected to experience negative pressures. It should be emphasized that negative pressures have not been observed in the Davenport system, however Figure 9.9 shows that the monitoring stations were not in the affected zones.

Table 9.5
Summary of Recorded and Calculated Pressure Values (in psi)

Site	Recorded Pressure	Case Study 1	Case Study 2	Case Study 3
1	77 – 43	77 – 40	77 – 23	77 - 40
2	50 – 42	50 – 44	33 – 27	50 – 44
3	50 – 9	55 – 3	52 – 3	55 – 12
4	62 - 50	72 – 48	66 – 39	72 – 47
5	- - - - *	55 – 50	51 – 46	55 – 50
6	78 – 63	82 – 55	78 – 56	82 – 55
7	70 - 50	87 - 20	62 - 25	87 - 20

* Flow to the monitor was shut off during this event and no recorded pressure data are available

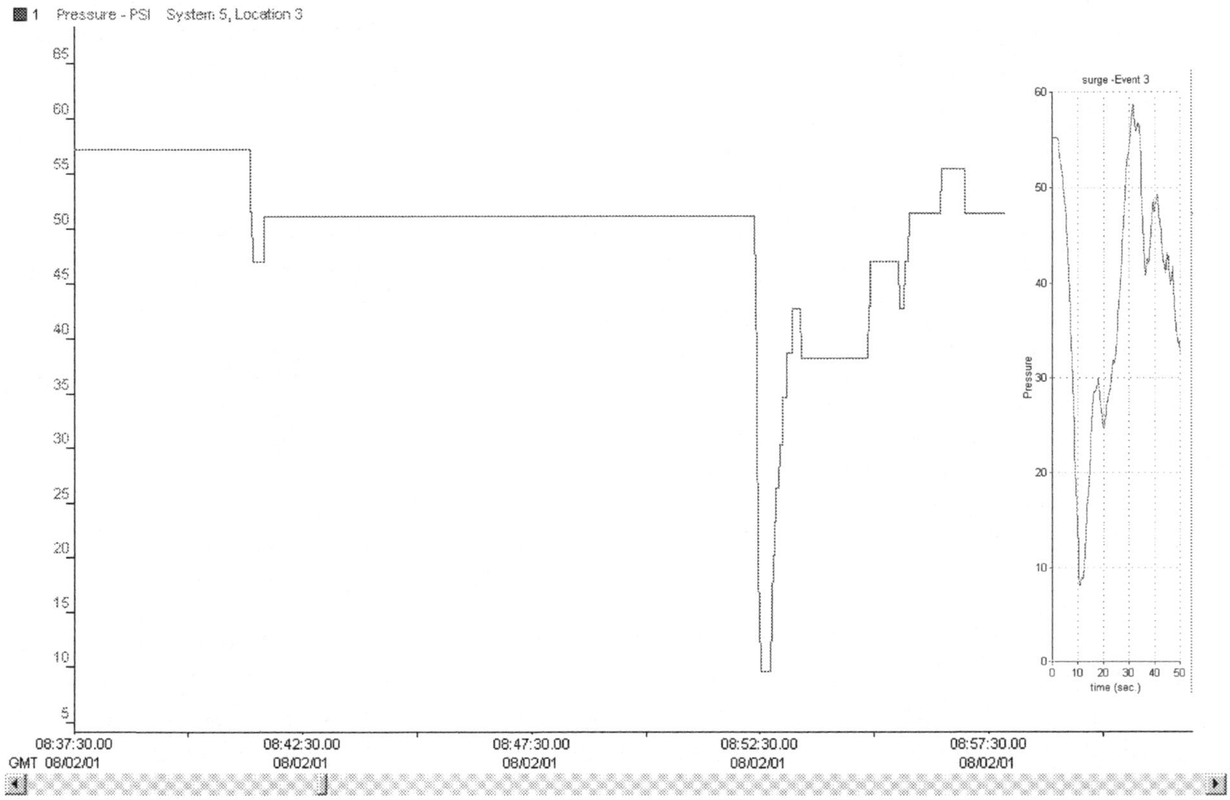

Figure 9.8. Pressure recording of the August 2 power outage (left) and the model simulation (right). The model well simulated the low-pressure transient.

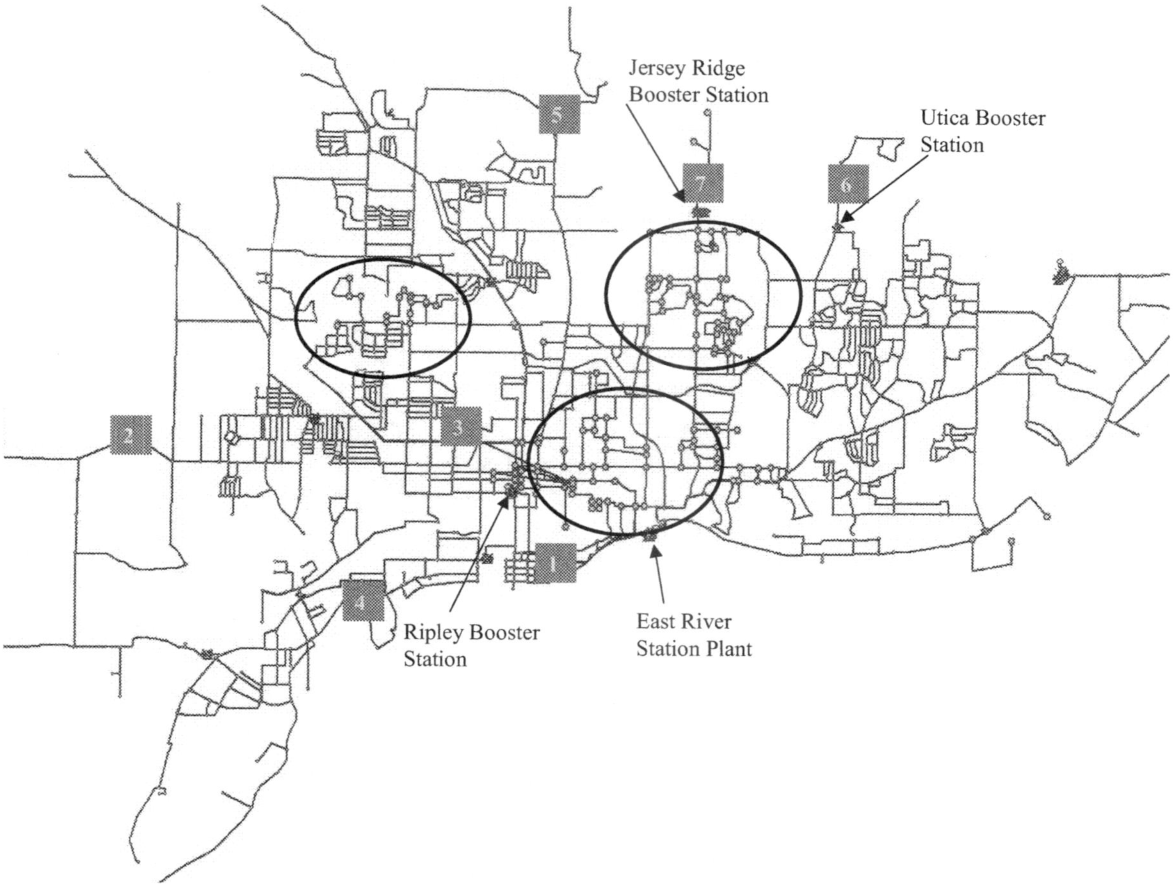

Figure 9.9. Model output showing negative pressure nodes (small circles) following the August 2 power outage (Case Study 1). The three zones most susceptible to low pressure are circled.

Following the completion of the study (June 2002) five of the pressure monitors (all except #5 and #7) were moved to fire hydrants within the circled areas in Figure 9.9. Monitoring of these locations continued until October 11, 2002 (when the potential for freezing required that the study be terminated) in an attempt to validate the model predictions. No negative pressure events occurred during June to October 2002.

A summary of power interruptions from June 2001 through May 2002 is presented in Table 9.6. Several interruptions were planned and a back up power generator was available to maintain pumping operations, however most outages were unplanned.

Table 9.6
Summary of Power Interruptions at the East Main Station Plant

Date	Power Time Off	Power Time Back On	Generator Came On	Scheduled Interruptible Power Outage
06/12/2001	6:59 PM	7:04 PM	No	No
06/12/2001	9:00 AM	9:05 AM	No	No
06/14/2001	6:12 PM	6:19 PM	No	No
06/18/2001	6:19 AM	6:25 AM	No	No
07/14/2001	7:14 AM	7:20 AM	No	No
07/21/2001	6:46 AM	6:57 AM	No	No
07/31/2001	12:58 PM	1:00 PM	12:51 PM	Yes
08/02/2001	3:41 PM	3:44 PM	No	No
08/02/2001	8:56 AM	9:02 AM	No	No
08/06/2001	2:13 PM	2:15 PM	1:56 PM	Yes
08/06/2001	6:51 AM	6:56 AM	No	No
08/07/2001	12:52 PM	12:54 PM	12:50 PM	Yes
11/09/2001	7:00AM	7:11AM	No	No
01/13/2002	08:53AM	08:53AM	N0	No
02/20/2002	07:40AM	07:40AM	No	No
03/09/2002	09:33AM	09:33AM	No	No
05/17/2002	02:20AM	02:20AM	No	No
05/22/2002	09:15PM	10:05PM	Yes	No

For case studies 4 and 6, where two surge tanks (closed 20-foot diameter tanks with 10 feet of water and 10 feet of air in the surge tank) were modeled downstream from the Main and High Service pumps stations (Figure 9.10), the computer output did not predict any negative pressure locations. Similarly, modeling scenarios where just the Utica Ridge or the Jersey Ridge pumps were tripped did not result in any predicted negative pressures. These case data (scenarios 1-6) indicate that the loss of power to the East River Station could result in significant distribution system low pressures (<0 psi). The number of pumps initially running and the level of demand appeared to have little effect on the generation of low pressures. It also appeared that power failures that do not affect the East River Station would generally not cause low pressures. The model predicts that installation of surge tanks downstream from the East River Station pumping station may be able to provide protection from negative pressure transients due to a power outage.

A modeling exercise was performed to evaluate the effectiveness of air/vacuum valves for reducing low pressures following a major pump trip event. Thirteen air/vacuum valves (4" orifice) were modeled at elevated locations in regions where low pressures were predicted. Figure 9.11 shows the locations of the simulated air/vacuum valves. The modeling scenario where all seven pumps were tripped (Case 4, Table 9. 4) was repeated with the air valves installed, however, the results (Figure 9.12) showed that the presence of the air/vacuum valves provided little protection. This result occurred even though the air/vacuum valves were activated as shown in pressure traces (Figure 9.12). One potential problem associated with the use of air/

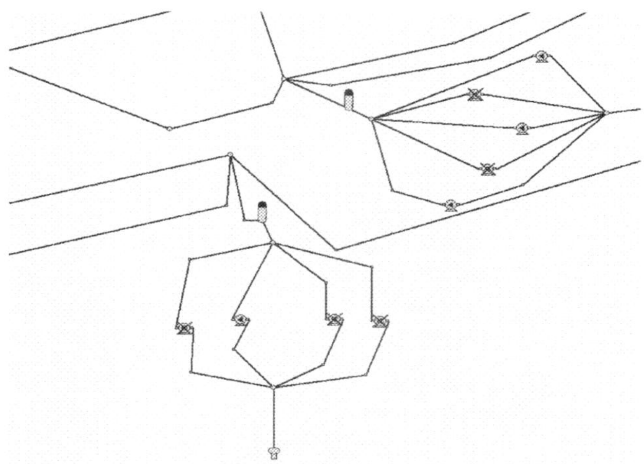

Figure 9.10. Location of simulated surge tanks (small columns) downstream of the nine Main and High Service pumps.

vacuum valves is an elevated surge pressure due to the "air slam" when air is expelled from the valve.

Two modeling scenarios were examined to simulate a 2-inch leak near to monitoring site 4 (Event #5, Table 9.2). As detailed previously, it was thought that a construction accident was the cause of a pressure drop at site 4 from 70 psi to around 20 psi. The model determined that a leak of 4,000 gpm would be required to produce the recorded drops in pressure – a flow rate that is excessive for a 2-inch line. Additionally, it was found that the more rapidly the leak developed (4 seconds compared to 50 seconds), the more the initial pressure dropped. Therefore, it is possible that a rapid puncture of a distribution system line may have resulted in the observed pressure variations, and that under certain circumstances such a rapid rupture could cause negative pressures at various distribution system locations. Figure 9.13 shows that the low-pressure locations for the 4-second scenario were down stream of the electronic pressure monitoring location

Modeling Additional Surge Scenarios

The Surge2000 model was used to examine several hypothetical scenarios related to hydrant use to examine the vulnerability of the Davenport distribution system to negative pressure transients. The purpose of these simulations was to determine if the operation of fire hydrants under emergency situations could contribute to negative pressure transients. It is important to note that there was no evidence of pressure events due to hydrant activity, but because the area of the distribution system affected by hydrant use may be limited, these activities may not have been observed at the fixed monitoring locations.

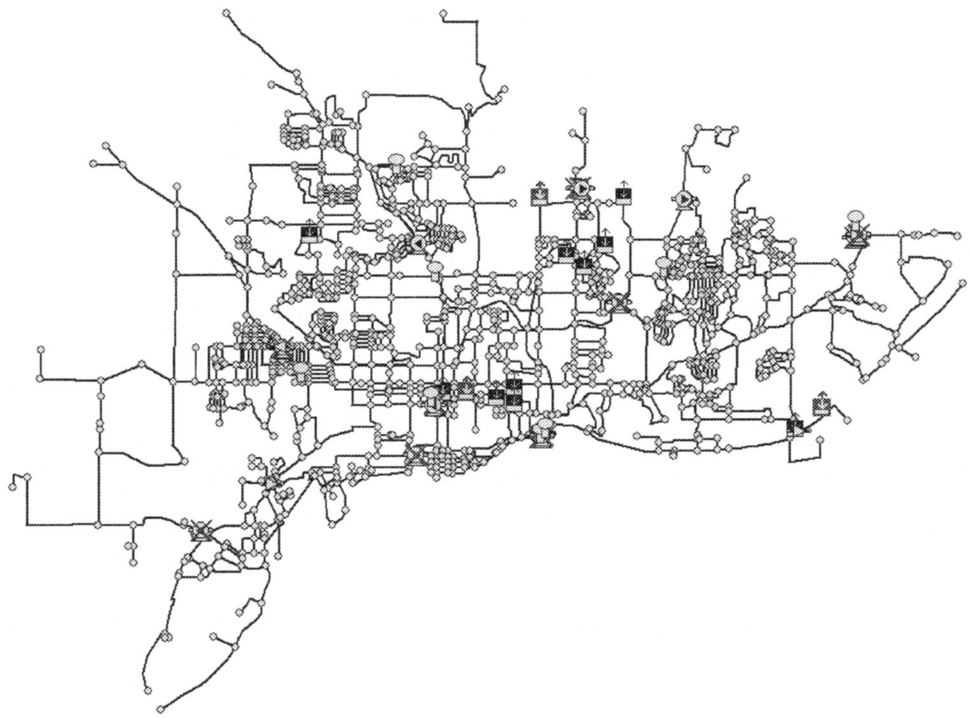

Figure 9.11. Locations (boxes) for simulated air/vacuum relief valves.

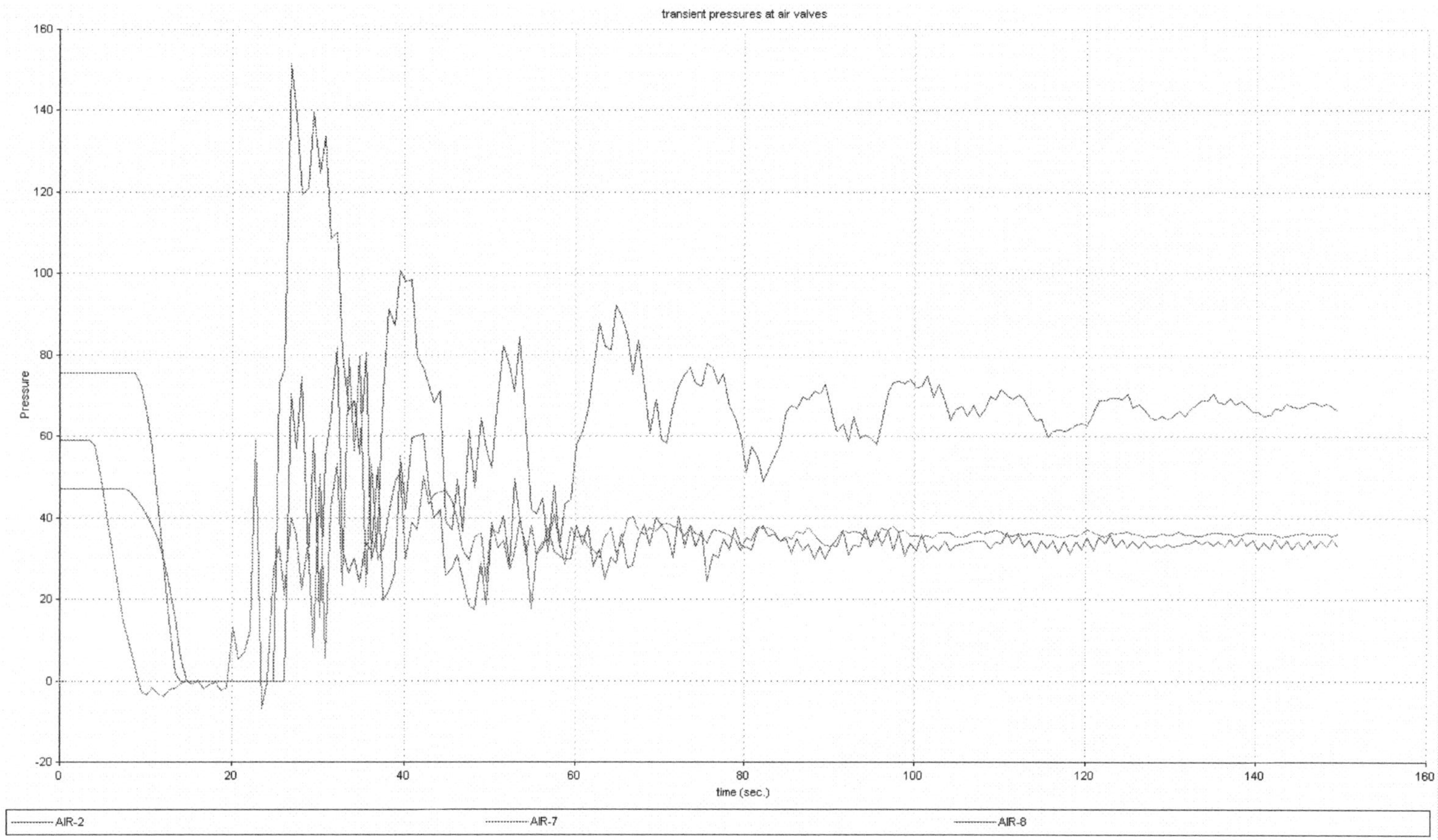

Figure 9.12. Pressure simulations following a 7-pump trip and insertion of air/vacuum relief valves. The valves failed to prevent the occurrence of negative pressure transients, but caused a spike in pressures due to "air slam" of the valves.

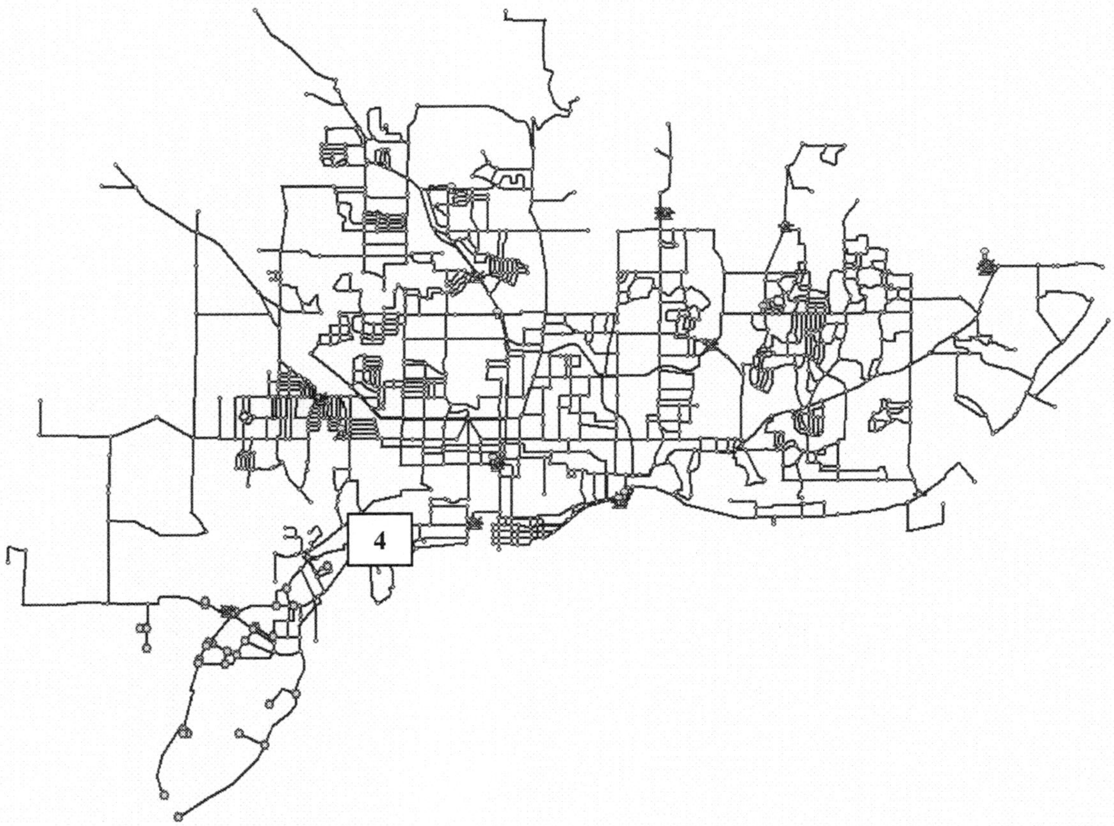

Figure 9.13. Model output showing negative pressure nodes (circles) following a 2-inch leak in zone 4. The model assumed that the leak developed within 4 seconds.

Event #5 illustrates the situation where a rapid demand in the system could cause negative pressures. Although this event was modeled as a pipe break, turning on a fire hydrant would produce the same response. If a fire pump was used to increase the fire flow, even lower pressures would result. The situation would be more severe when the hydrant is at a lower elevation than the adjacent regions of the distribution system. A simulated hydrant startup was modeled in a vulnerable area of the Davenport system. Figure 9.14 shows elevations in such a region. The hydrant location highlighted is in the lowest part of that region. A steady state calculation for a fire flow of 2400 gpm at the location will lower the pressure to 20 psi while the adjacent locations are at higher pressures.

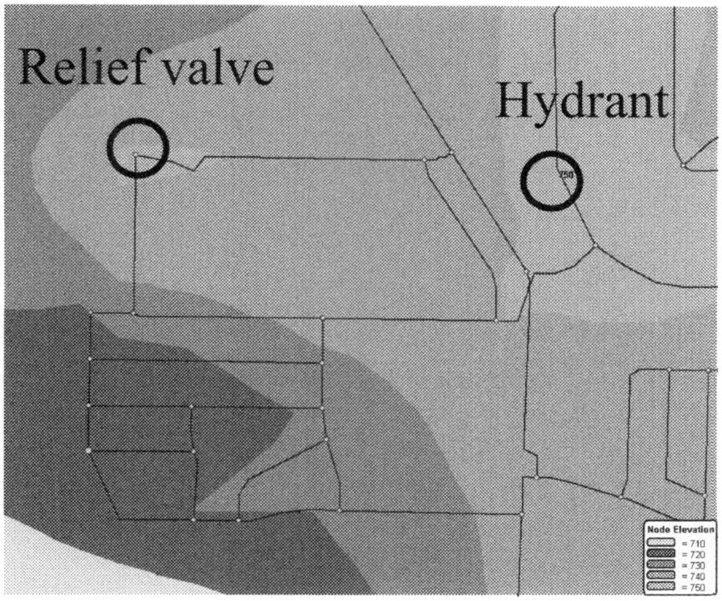

Figure 9.14. Elevations in a vulnerable area around a hydrant, Location of modeled air/vacuum relief

A surge analysis was performed for a 3 second and a 30 second hydrant startup. As shown in Table 9.7, a 3 second startup produces low pressures at 19 nodes in the model for several minutes. Assuming that these locations have leaks and are submerged within the water table, the model predicts the potential for intrusion of approximately 1 gallon (3.78 L) during the operation of the hydrant (Appendix H provides details on the calculation of intrusion). Figure 9.15 shows that the locations of the low pressures are predominately associated with the points of high elevation shown in Figure 9.14. Initiating the flow from the hydrant over a 30 second period dramatically reduced the number of nodes experiencing low pressures and the potential for intrusion (Table 9.7). An analysis was performed to evaluate the effectiveness of inserting an air/vacuum valve at a high spot in the vicinity of the hydrant startup (Figure 9.14). Data presented in Table 9.7 show that even though the air/vacuum valve was activated, it did little to reduce low pressures and the potential for intrusion in the region.

Table 9.7
Model Simulations of Hydrant Operations

Case [a]	Number of Nodes	Total Time (seconds)	Intrusion Integral [b]	Intrusion Volume (gal.)
Hydrant opening in 3 seconds	19	197	7.3	0.99
Hydrant opening in 30 seconds	1	3	0.1	0.01
Hydrant opening in 3 seconds with air valve	19	196	0.1	1

[a] Simulations were performed for a hydrant at model node 5975 at a flow of 2400 gpm.
[b] see Appendix H

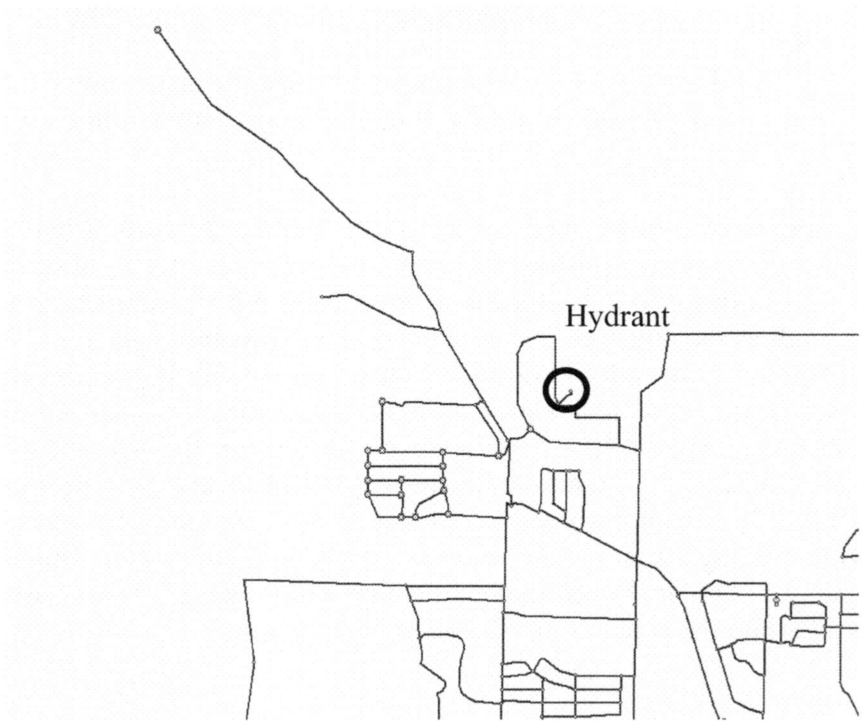

Figure 9.15. Locations of low pressure (circles) for the 3-second hydrant simulation.

CHAPTER 10
HOUSEHOLD RESULTS

Of the 456 households participating in the Water Evaluation Trial, water samples were collected from 107 participants to characterize their water quality. The study protocol required that the utility have no knowledge of which households participated in the study. Therefore, all samples were collected by the epidemiology team staff. This interaction limited the samples to three occasions: when the treatment devices (Figure 10.1) were first installed, at six-month crossover when the treatment devices were replaced, and at the 12-month completion of the study. For each occasion, a water sample was collected from the treatment unit and another was collected from the kitchen tap. The epidemiology team sample collectors were trained by Iowa-American personnel on the techniques for aseptic sample collection and review sessions were periodically performed to ensure proper technique, recalibrate equipment, and perform spilt sample analyses. All samples were placed on ice and transported within 6 hours to the Iowa-American laboratory for immediate analysis.

Figure 10.1. Treatment device with diverter valve detached.

Because of the limited access to the household sites, the objective of the monitoring was not to try to detect pathogenic microbes, but to characterize the overall water quality and compare these parameters to those from the treatment plant and distribution system to determine if significant changes had occurred in the microbial quality of water at the participants' taps or in the treatment devices.

COMPARISON OF TAP AND TREATMENT DEVICE PARAMETERS

A summary of the household tap and treatment device monitoring results is shown in Table 10.1. It is important to recall that, because of the "blinding" of the study, the device samples include both active (functional microfiltration and UV devices) and non-functional placebo devices. A total of 295 samples were collected for physiochemical and biological parameters including pH, temperature, turbidity, total chlorine, total coliform and HPC bacteria (one device sample was not collected). Overall the average pH (7.3) and turbidity (0.1 NTU) were similar between tap and device samples, and the ranges for these parameters were similar (Table 10.1). The average temperature of the device samples was slightly higher (11.3°C) than tap water samples (10.3°C), and total chlorine residuals were lower (1.62 mg/L for tap water samples and 1.53 mg/L for device samples). The difference in temperature and total chlorine residuals could be explained by the longer residence time of the water within the countertop treatment devices.

Coliform bacteria were detected in 64 out of 294 (21.8%) samples from the treatment devices (Table 10.2), with an average density of 18.4 cfu/100 mL. Coliform samples that were "too numerous to count" (TNTC) or showed confluent growth on the membrane filter were scored as containing greater than 200 cfu/100 mL. Twenty-two of the 64 (34.3%) coliform positive treatment device samples were scored as TNTC. All samples were transferred to an EC-MUG broth to test for the presence of *Escherichia coli*. One sample was *E. coli* positive. By comparison, only 10 of 295 (3.4%) tap water samples contained coliform bacteria (average 0.9 cfu/100 mL), and only one sample was TNTC (Table 10.3). None contained *E. coli*.

No water samples from the newly installed devices during the first six months of testing were positive for total coliform bacteria because these analyses evaluated new units (tested upon initial installation) (Figure 10.2). However, at the time of crossover (switching from active to placebo or *vice versa,* which began on June 4, 2001), following six months of continuous operation, 37 of 102 (36.3%) water samples were positive for total coliforms. The crossover process was completed by December 26, 2001. These new units were operated for another six-month period before starting to be removed in January 2002 (the final device was removed in June 2002). Coliform analyses of water samples collected from the cross-over replacement devices following six months of services resulted in 27 of 88 (30.7%) samples coliform positive. Coliforms isolated from the devices included *Klebsiella oxytoca, Enterobacter cloacae, Enterobacter agglomerans*, and *E. coli*. A review of sample collection procedures during these periods of high coliform occurrence was performed to ensure that the results were not due to sampling errors. No problems with sample collection procedures were detected.

Four units (2 active and 2 sham) were sent to Dr. Kim Fox, at the USEPA laboratory in Cincinnati for further examination (due to "blinding" the identity of the units, whether active or sham, was not known). A total of 10 samples were processed for each unit as listed in Table 10.4. In all cases, the swab or component was placed in P/A broth for 24 hours at 35°C and bacteria from turbid broth samples were identified. Coliform bacteria were detected in a total of eight samples from two devices (#5514 and 5508):

Device #5514	E, G
Device #5508	A, E, F, G, J, K

Table 10.1
Physiochemical and biological water quality parameters for the household treatment devices and taps.

Location	Parameters	N[a]	N < DL[b]	Mean	STD	Min	25%ile	50%ile	75%ile	90%ile	95%ile	Max
Treatment devices	pH (pH units)[c,d]	294		7.3	0.1	6.7	7.2	7.3	7.4	7.4	7.5	7.7
	Temperature (°C)[c,d]	294		11.27	5.46	4.00	7.00	9.00	15.00	20.00	22.00	25.00
	Total coliform (cfu/100mL)[c,d]	294	230	18.4	54.8	0	0	0	0	32	200	200
	Turbidity (ntu)[c,d]	294		0.11	0.06	0.04	0.08	0.10	0.12	0.15	0.17	0.90
	Total chlorine (mg/L)[c,d]	294		1.53	0.48	0.08	1.28	1.65	1.90	2.01	2.10	2.40
	HPC (cfu/mL)[c,d]	294	59	561.8	685.8	0.00	20	225	1080	1730	1940	2760
Taps	pH (pH units)[c]	295		7.3	0.1	6.6	7.2	7.3	7.4	7.5	7.5	8.1
	Temperature (°C)[c]	295		10.31	5.63	3.00	6.00	8.00	15.00	20.00	20.00	25.00
	Total coliform (cfu/100mL)[c]	295	285	0.9	11.9	0	0	0	0	0	0	200
	Turbidity (ntu)[c]	295		0.1	0.05	0.07	0.10	0.12	0.14	0.16	0.18	0.80
	Total chlorine (mg/L)[c]	295		1.62	0.45	0.08	1.38	1.76	1.90	2.07	2.19	2.40
	HPC (cfu/mL)[c]	295	53	140.7	325.2	0.	10	30	90	380	900	1940

[a] Number of samples [b] Number of samples below the detection limit i.e. not detected. [c] Household sampling was not performed in December 2001.
[d] One sample was not collected.

All others samples were negative for coliforms. On unit 5514, two sample sites (E; rubber o-ring and G; aeration screen, Figure 10.3) were positive for coliforms and in both cases the coliform was identified as *Enterobacter* (*Pantoea*) *agglomerans*. On unit 5508, six sample sites (A, E, F, G, J, K, see Table 10.4) were positive for coliforms and all were identified as *Enterobacter cloacae*. The detection of coliform bacteria on the UV glass suggests that this unit was an inactive sham, but it is possible that water contained in the unit may have flowed back from the valve assembly into the UV chamber during shipping. The isolation of a single coliform species from each unit suggests that the bacteria colonized and grew within the units.

Evidence for bacterial growth in the treatment devices was also provided by examining the HPC data. For tap water samples the average level of HPC bacteria was 140.7 cfu/100 mL (range 0 to 1,940 cfu/100 mL), whereas levels in the device samples averaged four times higher (average 561.8 cfu/100 mL, range 0 to 2,760 cfu/100 mL).

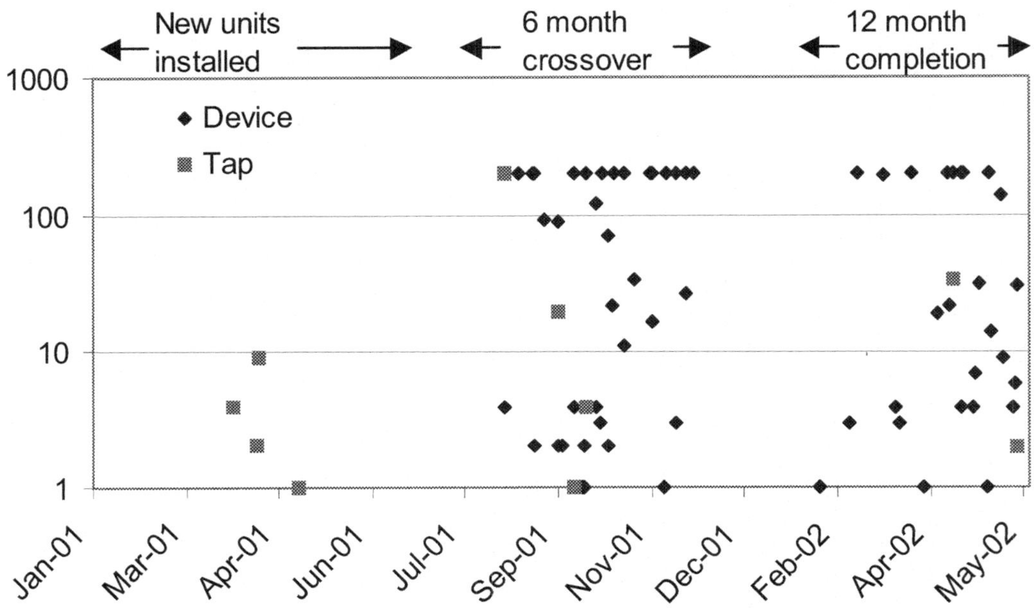

Figure 10.2. Occurrence of Coliform Bacteria in Tap and Device Samples

Table 10.2
Total coliform for household treatment devices (sorted by descending CFU/100mL)

Sample date	CFU /100 mL
10/11/01	>200
10/18/01	>200
10/23/01	>200
10/3/01	>200
10/3/01	>200
11/15/01	>200
11/20/01	>200
11/26/01	>200
11/30/01	>200
11/6/01	>200
11/7/01	>200
2/26/02	>200
3/27/02	>200
4/15/02	>200
4/18/02	>200
4/22/02	>200
4/24/02	>200
5/8/02	>200
8/27/01	>200
9/27/01	>200
9/4/01	>200
9/5/01	>200
3/12/02	193
5/14/02	137
10/8/01	122
9/10/01	91
9/18/01	90
10/15/01	70
10/29/01	34
5/2/02	32
5/23/02	31
11/26/01	27
10/17/01	22
4/16/02	22
4/10/02	19
11/7/01	17
5/9/02	14
10/23/01	11
5/15/02	9
4/30/02	7
5/21/02	6
10/8/01	4
3/18/02	4
4/22/02	4

Sample date	CFU/100 mL
4/29/02	4
5/20/02	4
8/20/01	4
9/27/01	4
10/10/01	3
11/20/01	3
2/21/02	3
3/20/02	3
10/15/01	2
10/2/01	2
9/18/01	2
9/20/01	2
9/5/01	2
10/1/01	1
10/2/01	1
11/14/01	1
2/5/02	1
4/2/02	1
4/2/02	1
5/7/02	1

Table 10.3
Total coliform from household tap samples (sorted by descending CFU/100 mL)

Sample date	CFU/100 mL
8/20/01	>200
4/18/02	34
9/18/01	20
4/10/01	9
10/3/01	4
3/27/01	4
4/9/01	2
5/23/02	2
5/2/01	1
9/27/01	1

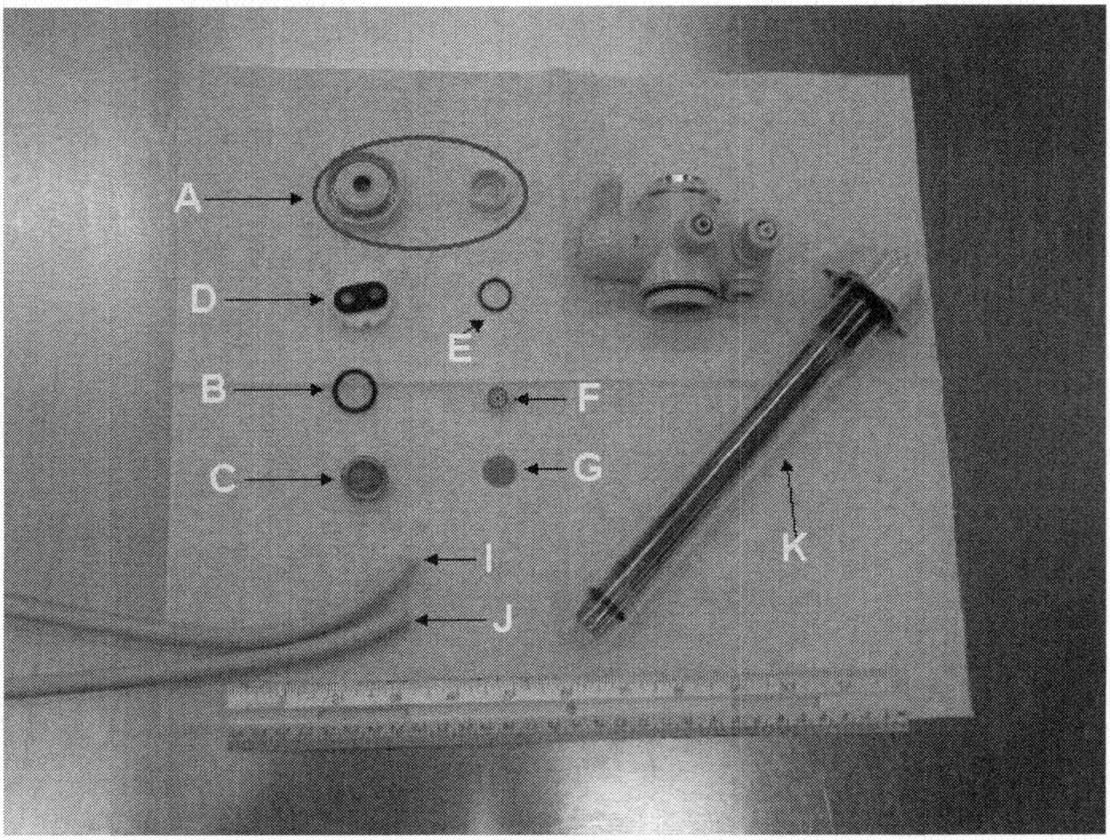

Figure 10.3. Filter device samples

Table 10.4
Treatment device samples

Code	Location
A	The outside plastic housing from diverter (swab material)
B	Rubber o-ring from faucet side of diverter
C	Aerator screen from faucet side of diverter unit
D	Flow controlling parts inside of diverter
E	Rubber o-ring from treated water side of diverter
F	Plastic aerator from treated side of diverter
G	Aeration screen from treated side of diverter
I	Inside of tap water line (swab)
J	Inside of treated water line (swab)
K	U.V. glass (swab)

COMPARISON OF TAP WATER, DISTRIBUTION SYSTEM, AND PLANT EFFLUENT PARAMETERS

Comparison of data from the plant effluent (Chapter 7), distribution system (Chapter 8), and household taps (Chapter 10) allows an assessment of the magnitude of water quality changes. A substantial difference in water quality from plant effluent to the consumer tap would be indicative of contamination at some point in the delivery process. Water quality variables that were common to the three locations included: pH, temperature, total coliform, HPC bacteria, total chlorine, and turbidity. Table 10.5 provides a summary of average values for each parameter and location.

Table 10.5
Average water quality values for plant effluent, distribution system, and household tap water

Parameter	Plant Effluent		Distribution System		Household Tap [b]	
	N[a]	Mean	N[a]	Mean	N[a]	Mean
pH (pH units)	577	7.4	358	7.4	295	7.3
Temperature (°C) [c]	577	12.08	358	13.63	295	10.31
Total coliform (cfu/100mL)	408	0.0	2471	0.02	295	0.9
Turbidity (ntu)	577	0.052	394	0.14	295	0.1
Total chlorine (mg/L)	577	2.54	2471	2.08	295	1.62
HPC (cfu/mL)	376	55.7	2430	231.9	295	140.7

[a] Number of samples. [b] Household sampling was not performed in December 2001.
[c] Temperature was not recorded for plant effluent; raw water values are provided.

Figures 10.4 through 10.9 show box-and-whisker diagrams providing a graphical summary of the environmental factors datasets collected from the three sampling locations. In these diagrams the box outlines the 25th and 75th percentile of the data, and the line in the middle of the box shows the median (50th percentile). The "whiskers" are formed by the minimum and maximum values. Emphasis should be placed on the 25th, 50th, and 75th percentile data, as the whiskers highlight a single point out of hundreds or thousands values collected.

In general, the degree of variability (the length of the whisker) increased as the water traveled from the plant effluent to the consumer's tap. There was no statistical difference (using the non-parametric Wilcoxon Rank Sum test for non-normally distributed data) between plant effluent and distribution system pH data (Table 10.6), and the small change in household tap pH (0.1 pH units) was not substantial (Figure 10.4). Surprisingly, the average water temperature for the household data was lower than the distribution system value (Figure 10.5), but these results are attributed to the fact that the household samples were collected predominately in the winter months (beginning in January 2001). In addition, because the taps were flushed for 3 to 5 minutes before sample collection, warm water that might have been within the household plumbing was not collected.

There was no statistical difference ($p=0.566$) in total coliform levels between plant effluent and distribution system samples (Table 10.5), but tap water coliform levels increased to 0.9 cfu/100 mL (Figure 10.6). However, the overall frequency of coliform occurrence was low, with none found in the plant effluent, 0.08% (2 of 2471) in the distribution system, and 3.4% (9 of 295) in the household tap samples. The increase in the tap coliform levels can be attributed to bacterial regrowth and contamination of the faucets by household activities (food preparation, dish washing, etc.) in the kitchen sink. Most of the coliform genera detected (*Klebsiella*, *Enterobacter*) are known to be associated with coliform regrowth events (LeChevallier 1990). It should be emphasized that Iowa Department of Natural Resource criteria for collection of coliform samples (dual handle, short neck faucet) could not be applied to most kitchen sink locations (most were single handle, long neck). Sampling personnel reported that the sanitary condition of some locations was quite poor. Alternatively it is possible that cross connections within the home plumbing could be responsible for the increased household coliform results. An additional USEPA study is underway to address this question, and initial results have detected some cross connections of sanitary significance in several of the participating households (P. Schwartz, personal communication, 2002).

Turbidity values were significantly higher ($p<0.01$) in the distribution system and household tap samples than in the treatment plant effluent (Table 10.6). Moreover, the household tap turbidity data were significantly lower than the distribution system values. The presence of corrosion products and post-treatment chemical precipitation could account for the turbidity increase in the household and distribution system samples. It is unclear why the household turbidity data should be lower than the distribution system values, although the range of values was large (Figure 10.7).

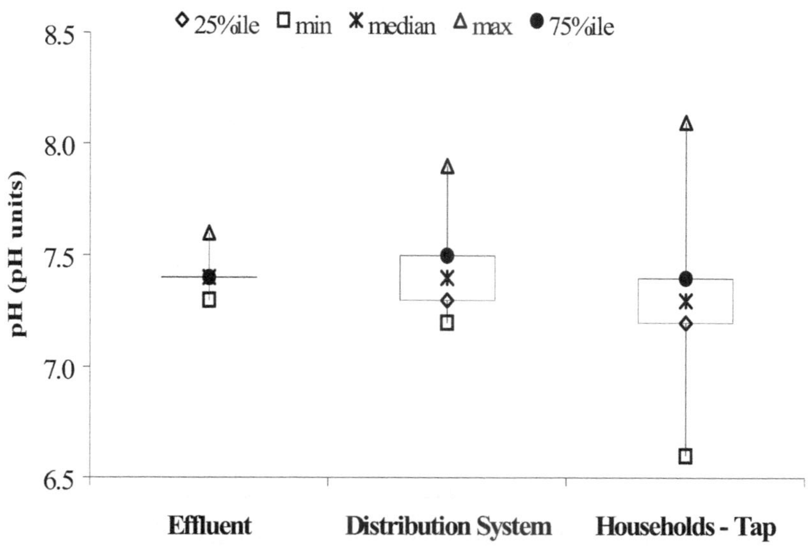

Figure 10.4. pH data from plant effluent, distribution system, and household tap water.

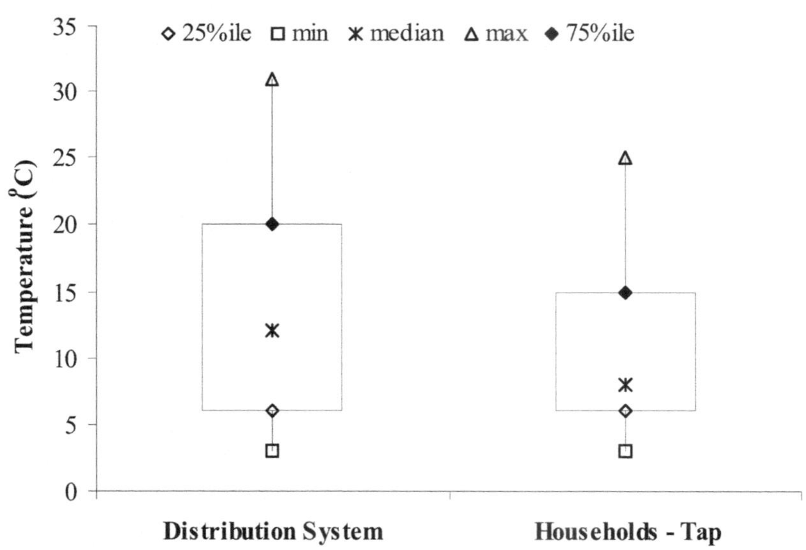

Figure 10.5. Temperature data from distribution system and household tap water. Temperature was not measured in the treatment plant effluent.

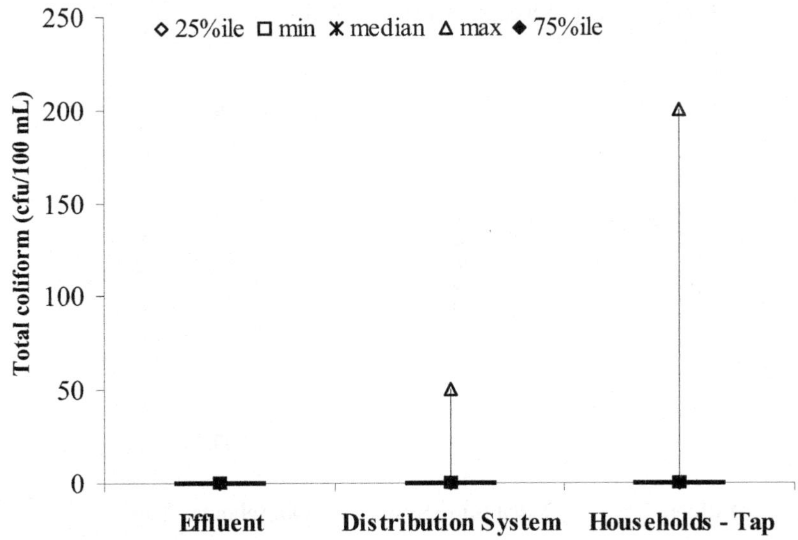

Figure 10.6. Total coliform data from plant effluent, distribution system, and household tap samples.

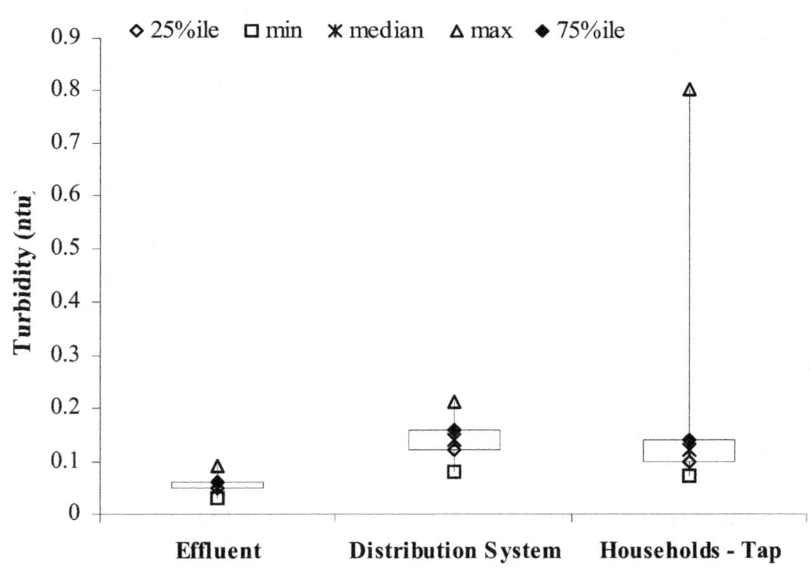

Figure 10.7. Turbidity data from plant effluent, distribution system, and household tap water.

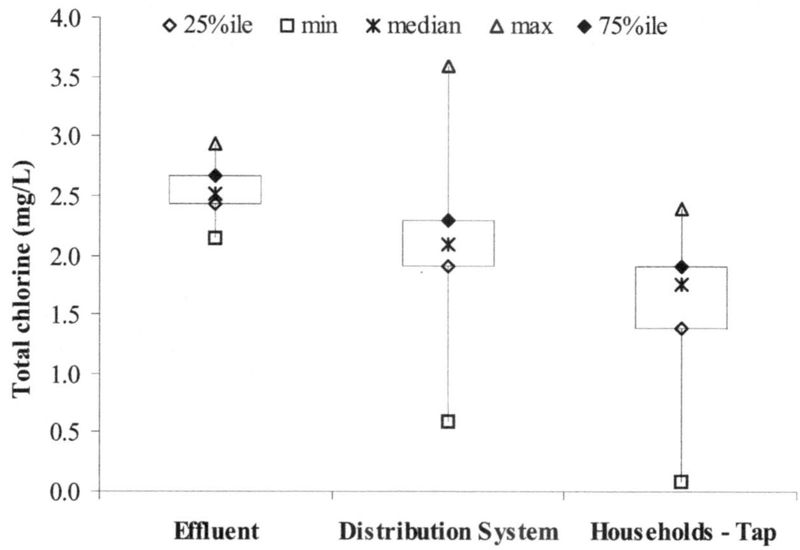

Figure 10.8. Total chlorine data from plant effluent, distribution system, and household tap samples.

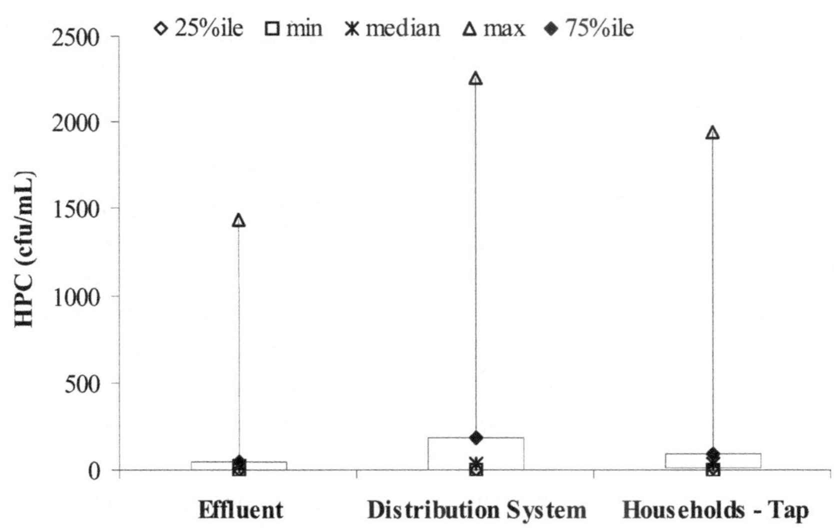

Figure 10.9. HPC data from plant effluent, distribution system, and household tap water.

Table 10.6.

Wilcoxon rank sum comparison of plant effluent, distribution system, and household tap samples

Parameter	Location 1	Parameter 2	Mean Rank Sum Location 1	Mean Rank Sum Location 2	P-value	Statistically different?
pH	Effluent	Distribution System	478.5	451.0	0.106	No
	Distribution System	Household tap	376.6	266.8	<0.010	Yes
	Effluent	Household tap	505.6	301.4	<0.010	Yes
Temperature	Distribution System	Household tap	351.6	297.1	<0.010	Yes
Total coliform	Effluent	Distribution System	1439.0	1440.16	0.566	No
	Distribution System	Household tap	1378.6	1424.4	<0.010	Yes
	Effluent	Household tap	347.0	358.9	<0.010	Yes
Turbidity	Effluent	Distribution System	289.07	774.40	<0.010	Yes
	Distribution System	Household tap	405.59	264.08	<0.010	Yes
	Effluent	Household tap	289.49	724.05	<0.010	Yes
Total chlorine	Effluent	Distribution System	2509.44	1294.51	<0.010	Yes
	Distribution System	Household tap	1473.44	630.1	<0.010	Yes
	Effluent	Household tap	582.96	150.04	<0.010	Yes
HPC	Effluent	Distribution System	1014.29	1463.72	<0.010	Yes
	Distribution System	Household tap	1382.84	1199.57	<0.010	Yes
	Effluent	Household tap	305.51	374.86	<0.010	Yes

* Mean rank sum = sum of the ranks is normalized based on the sample size n (sum of ranks/n)

Total chlorine levels significantly declined from the plant effluent, to the distribution system, to the household tap samples (Figure 10.8). All plant effluent total chlorine residuals were greater than 2.0 mg/L, and all distribution system values were greater than 0.5 mg/L. Despite one low household tap total chlorine residual (0.08 mg/L), most values were greater than 1.0 mg/L. Nine of the 10 positive household tap coliform samples had total chlorine residuals greater than 1.0 mg/L. The one household tap TNTC coliform sample had a total chlorine residual of 2.35 mg/L. The presence of coliform bacteria in samples with high total chlorine residuals suggests that the origin of the bacteria was close to the point sample collection.

The significant difference in HPC bacterial densities between the plant effluent, the distribution system, and household tap samples (Figure 10.9) demonstrated the potential for bacterial growth in the water supply. Similar to the coliform data, the lack of correlation between HPC bacterial levels and total chlorine residuals (Figure 10.10) suggest that the origin of the bacteria was close to the point of sampling. The results for the tap water total coliform and HPC data, and the coliform bacteria in the treatment devices all suggest that the bacterial levels observed were primarily due to bacterial regrowth.

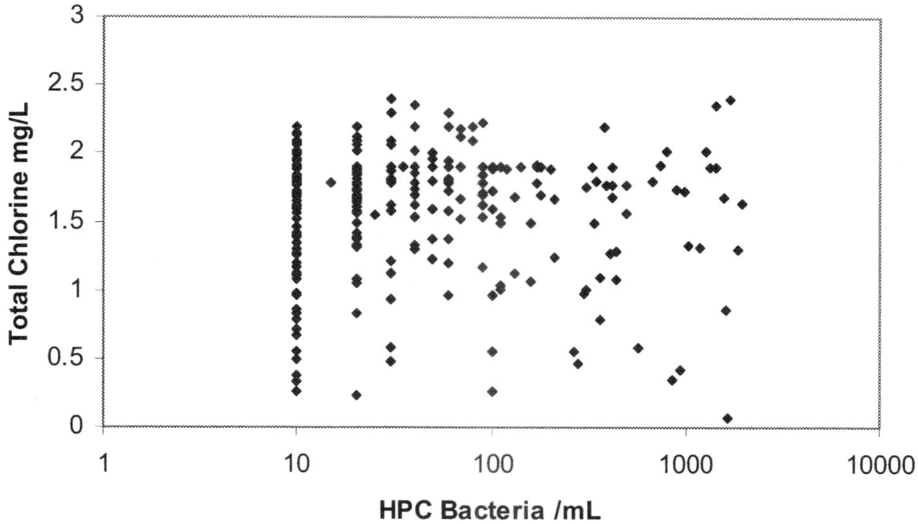

Figure 10.10. Relationship between total chlorine residual and HPC bacteria in household tap samples.

CHAPTER 11
DISCUSSION, CONCLUSIONS AND RECOMMENDATIONS

The primary objective of the project was to characterize raw, treatment plant, distribution system, and household tap water quality with respect to microbes, water quality indicators, and physicochemical parameters so as to provide a context for interpreting the public health data generated by the epidemiology study. These data have been presented in the preceding chapters and are summarized below.

RAW WATER DATA

The Davenport, Iowa site was selected in part because the raw water source was "microbially challenged," and monitoring data collected by this study and a previous AwwaRF study (#488) documented the level of *Giardia*, *Cryptosporidium*, cultural enteroviruses, and other indicators of fecal pollution. Using Method 1623, *Giardia* cysts were detected in 31.1% of the weekly samples with an average of 0.093 cysts/L, and *Cryptosporidium* oocysts were detected in 13.5% of the samples with an average of 0.019 oocysts/L. Infectious *Cryptosporidium* oocysts were observed in 2.8% of the samples using the CC-PCR method. Culturable viruses were detected in 57.9% of the monthly raw water samples, with an average 6.72 MPN/100 L. The occurrence and concentration of these pathogens increased during the colder months and they were generally absent during the summer. The Mississippi River experienced a flood between April 14 and May 23, 2001, and increased concentrations were observed during this period for *Giardia*, male-specific coliphage, and total and fecal coliforms. Comparison of microbial results with the database developed during a previous AwwaRF project (LeChevallier et a. 2002) showed that the source water quality parameters measured during the study were fairly representative of the Mississippi River at Davenport, Iowa. It is interesting to note that despite the "microbially challenged" nature of the Mississippi River, there is only a 20 to 38% probability that the raw water *Cryptosporidium* data will exceed the 0.075 oocyst/L bin threshold required by the LT2ESWTR (USEPA 2000).

TREATMENT PLANT DATA

Extensive monitoring of the East River Station in Davenport, IA was conducted to document the plant performance for control of microbial pathogens and the production of consistent water quality. All the data indicated that the plant met or surpassed all drinking water criteria. No organisms of public health significance were detected in the treated water. The treatment plant was uniformly negative for all measurements of infectious *Cryptosporidium*, culturable enteric viruses, total coliforms, somatic coliphage, and *Clostridium perfringens* spores samples. Measurement of plant performance ranged from 1.5 to >6.36 log removal of microbes or particulate material, with the observed level of plant performance generally limited by the raw water concentration. The average plant effluent turbidity was 0.05 NTU, and the average 20 individual filter effluent turbidity values ranged from 0.034 to 0.055 NTU. Of the 10,991 daily individual filter turbidity values recorded, only four exceeded 0.15 NTU and none was greater than 0.45 NTU. The level of disinfection averaged 7 times greater than the minimum required for *Giardia* inactivation.

Despite raw water flood conditions between April 14 and May 23, 2001 finished water quality remained excellent with individual filter turbidities and combined filter effluent turbidities consistently less than 0.10 NTU. Filtered water particle counts were low, averaging less than 18 particles/mL, and free chlorine CxT values averaging 3.5 times higher than the required level of *Giardia* inactivation.

DISTRIBUTION SYSTEM WATER QUALITY DATA

Because the Iowa-American Water Company Davenport district distribution system includes more than 541 miles of mains within seven pressure zones, it was not possible to analyze all parts of the system with sufficient frequency to detect transient contamination events. Nonetheless, by all conventional measures of water quality, the drinking water met or surpassed all standards. Over 99.9% of the distribution system samples were free of total coliform bacteria, and 100% were free of *E. coli*. The average distribution system total chlorine residual was 2.08 mg/L and the minimum total chlorine level detected (based on 2,471 samples) was 0.59 mg/L. The average HPC bacterial level was 230 cfu/mL and some evidence of bacterial regrowth was apparent within the distribution system. Previous studies measured average AOC levels ranging from 74 to 78 μg/L in the treated effluent of the East River Station (Volk et al. 1996, Volk et al. 1997).

The low level occurrence (average concentration of 0.1 pfu/100 mL) of male-specific coliphage detected in 5.6% of the distribution samples may be associated with frequent main breaks during the winter months. However, the association between main breaks and the detection of coliphage was not particularly strong and additional research will be necessary using GIS-linked mapping to establish further relationships between these parameters. It is unlikely that male-specific coliphage could emanate from biofilm bacteria because bacterial pilli are not produced at cold temperatures. The detection of serotype 1 phage suggests that the source of the phage was not human fecal material, but the resistance of coliphage to chloramine disinfection (Hoff 1986) and survival at low water temperatures could serve to promote prolonged survival in drinking water. Additional studies of coliphage occurrence and survival in drinking water should be conducted to better assess the potential for using these organisms as indicators of distribution system contamination.

PRESSURE MONITORING AND MODELING

Detailed monitoring of distribution system pressure and hydraulic modeling of the pipe network was performed to supplement the microbial monitoring. Extensive pressure monitoring of the Davenport distribution system using seven electronic data loggers, collecting a combined 3,190 days of pressure monitoring totaling 2.756×10^8 data points, found only five occasions when distribution system pressures were less than 20 psi. No negative pressures were observed. Although the cause of these low-pressure events was not fully known, it was thought that most events were caused by pump shutdowns. This conclusion is supported by modeling data that suggests that significant low pressures ([0 psi) were predicted when the East River pumping station loses power. The number of operating pumps and the pumping rate appeared to have little effect on the generation of low pressures. Power failures that do not affect the East River station are thought not to cause low distribution system pressures. The model simulations also suggested that protection against low pressures due to power outages at East River was possible by installing surge tanks downstream of the pumping stations. Moreover, air/vacuum valves did not appear to offer reliable protection against low pressures in the Davenport system. It is recommended that Iowa-American consider the installation of surge tanks within its next engineering review (which is currently being performed). Additional monitoring using electronic pressure recorders is recommended to better assess the potential for negative pressure transients within the Davenport system. Special emphasis should be placed on the locations where the model predicts low pressures are possible. To facilitate monitoring, these recorders should be set to record and send an alarm to the control center only when low pressures are encountered. This design would alleviate the need to download data from the recorders on a weekly basis as was required for this study. Finally, distribution system operators should be aware that main breaks (which result in high flows), and the rapid opening of certain fire hydrants could result in low distribution system pressures. Training and the development of appropriate standard operating procedures to respond to such situations would be beneficial in maintaining a high level of distribution system water quality.

HOUSEHOLD TAP AND TREATMENT DEVICE DATA

Because of the limited access to the household sites, the objective of the household monitoring was to characterize the overall water quality of the treatment devices and kitchen taps, and to compare the data to results from the treatment plant and distribution system to determine if significant changes had occurred in the microbial quality of water. Overall, the values for pH, turbidity, temperature, and total chlorine residual were not substantially different between the tap and device samples. However, the detection of total coliform bacteria was 6.5 times more frequent, and levels were 20 times higher in the treatment device samples than in the normal tap samples. Analysis of several treatment devices detected total coliform bacteria within the diverter valve and in the UV treatment unit, suggesting that the bacteria were growing within the devices. The four-fold higher levels of HPC bacteria in the devices (compared to the taps) also supports the regrowth hypothesis. The species of total coliform bacteria identified in the units and tap water samples are typically associated with regrowth in distribution system biofilms (LeChevallier 1990, Geldreich and LeChevallier 1999). The detection of one *E. coli* positive device sample suggests that the coliforms may have originated from contamination of the faucets by household activities (food preparation, dish washing, etc.) in the kitchen sink. An alternative possibility that is currently being examined is that some of the contamination may have resulted from cross connections within the household plumbing. There is little evidence to suggest that the organisms originated from the treatment plant or the distribution system. Although comparison of water quality data from the treatment plant, distribution system, and household taps showed the expected increase in HPC bacteria and decrease in total chlorine residuals, all of the parameters still surpassed acceptable water quality standards. Only the treatment units failed to meet total coliform criteria (the 21.8% occurrence rate exceeded the 5% acceptable level), although it should be noted that the water samples from the treatment devices could not be collected using procedures conforming to bacteriological requirements (USEPA 1989a, APHA et al. 1998). For compliance purposes, total coliform samples cannot be collected from taps that have any attachments such as aerator screens or diverter valves. The faucets must be flushed for 3-5 minutes until water "representative of the distribution system" is sampled. These requirements are intended to avoid collection of total coliform samples that may be influenced by factors that could result in bacterial regrowth at the tap. These conditions could not be avoided in this study.

Nevertheless, the occurrence of high levels of total coliform bacteria, and one *E. coli* positive sample, raises the possibility that the treatment units themselves could result in elevated gastrointestinal risk for the study participants. This could be true if the inactive (placebo) units had higher occurrence rates, or concentrations of coliforms, than the active devices. Because of the "blinding" of the study this question could not be fully explored at this time. This assessment should be more fully explored once the epidemiology study results are revealed. To assess whether the units may be responsible for gastrointestinal disease, the researchers should examine the illness rates of the study participants for the months just prior to, and after, the treatment device crossover. The total coliform and HPC bacterial levels should be linked to their respective households to determine if households with high microbial levels had different levels of gastrointestinal disease than families with uncontaminated units. In addition, because new units were installed at the crossover, illness rates should drop if the bacterial levels were linked to disease. The epidemiology team performed a preliminary analysis of raw illness rates during the study to determine if the units were an unintentional source of illness (there was no evidence of increased rates at the time), but a more complete analysis should be performed once the study is completed.

SUMMARY

The extensive monitoring of the Davenport system failed to find significant indicators of microbial risk. The drinking water in the Davenport system met or surpassed all measures of water quality. Despite using a "microbially challenged" source water, the operators of the East River Station met the challenge by consistently producing a high quality product. The resiliency of the treatment process was evident during the Spring 2001 flood when, despite degraded raw water quality, finished drinking water quality remained outstanding.

The study did identify some areas for future consideration and research, including the relationship between the occurrence of male-specific coliphage and distribution system main breaks, the potential for negative pressure transients due to power outages, and the need for more comprehensive distribution system pressure monitoring. However, there was no clear link between these events and the degradation of drinking water quality.

The study also provided some useful direction to resolve some of these issues. Installation of surge control tanks downstream of the high service pumps was modeled to eliminate the potential for negative pressure transients due to power outages. Distribution system models were shown to be useful to guide the placement of electronic monitors in areas that may be most vulnerable to low pressure events. Electronic reporting of low-pressure events could better alert treatment plant operators to events within the treatment process that may trigger water quality problems within the distribution system. Coliphage monitoring, especially in chloraminated systems, may be a sensitive indicator of contamination due to cold-weather main breaks. Further research is necessary.

Finally, the microbial benefits of point-of-use treatment devices remains in doubt. The levels of HPC and total coliform bacteria found within the treated effluent raise concerns that units may be generators of microbial risk. The design of this study was not focused to address this concern, but analysis of epidemiological data may be useful in providing some insights. Additional studies are needed to examine the pathways and the types of microbes (*Legionella*, *Mycobacterium*, enteric pathogens, etc.) that may contaminate point-of-use devices in the kitchen environment. Studies of household cross connections should also prove useful in addressing the total microbial risk that can be transmitted through water.

CONCLUSIONS

Raw Water

- Using Method 1623, *Giardia* cysts were detected in 31.1% of the weekly samples with an average of 0.093 cysts/L, and *Cryptosporidium* oocysts were detected in 13.5% of the samples with an average of 0.019 oocysts/L.
- Infectious *Cryptosporidium* oocysts were observed in 2.8% of the samples using the CC-PCR method.
- Culturable viruses were detected in 57.9% of the monthly raw water samples, with an average 6.72 MPN/100 L.
- The occurrence and concentration of pathogens was seasonally distributed with the highest levels during the colder months and were generally absent during the summer.
- The Mississippi River experienced a flood between April 14 and May 23, 2001, and increased concentrations were observed during this period for *Giardia* cysts, male-specific coliphage, and total and fecal coliforms.
- Comparison of microbial results with previous databases showed that the source water quality parameters were representative of the Mississippi River at Davenport, Iowa.

Treatment Plant

- The East River Station met or surpassed all drinking water compliance criteria. No organisms of public health significance were detected in the treated water.
- The treatment plant results were uniformly negative for detection of infectious *Cryptosporidium*, culturable enteric viruses, total coliforms, somatic coliphage, and *Clostridium perfringens* spores.
- Measurement of plant performance ranged from 1.5 to >6.36 log removal of microbes or particulate material, with the observed level of plant performance generally limited by the raw water concentration.
- The average plant effluent turbidity was 0.05 NTU, and the average turbidity of the 20 individual filter effluents ranged from 0.034 to 0.055 NTU. Of the 10,991 daily individual filter effluent turbidity values recorded, only four exceeded 0.15 NTU and none was greater that 0.45 NTU.
- The level of disinfection averaged 7 times greater than the minimum required for *Giardia* inactivation.
- Despite raw water flood conditions between April 14 and May 23, 2001 finished water quality remained excellent with individual filter turbidities and combined filter effluent turbidities consistently less than 0.10 NTU, filtered water particle counts were low, and free chlorine CxT values were 3.5 times higher than the required level of *Giardia* inactivation.

Distribution System

- Over 99.9% of the distribution system samples were free of total coliform bacteria, and 100% of the samples were free of *E. coli*.
- The average distribution system total chlorine residual was 2.08 mg/L and the minimum total chlorine level detected was 0.59 mg/L.

- The average HPC bacterial level was 230 cfu/mL and some evidence of bacterial regrowth was apparent within the distribution system.
- Low levels of male-specific coliphage (average concentration of 0.1 pfu/100 mL) were detected in 5.6% of the distribution samples.
- The detection of serotype 1 phage suggests that the source of the virus was not human fecal material.

Pressure Monitoring and Modeling

- Distribution system pressures less than 20 psi were observed on only five occasions, and no negative pressures were observed.
- Modeling data predicts that significant low-pressure ([0 psi) events occur only when the East River pumping station loses power.
- The modeling data suggest that power failures that do not affect the East River station would generally not cause distribution system low pressures.
- It is possible, based on model predictions, to provide protection against low pressures due to power outages at the East River Station by installing surge tanks downstream of the pumping stations.
- Modeling did not predict that air/vacuum valves would offer reliable protection against low pressures in the Davenport system.

Household Tap and Device Samples

- Values for pH, turbidity, temperature, and total chlorine residual were not substantially different between household tap and device samples.
- The detection of total coliform bacteria was 6.5 times more frequent, and levels were 20 times higher in the treatment device samples than in the tap samples.
- Analysis of several treatment devices detected total coliform bacteria within the diverter valve and in the UV treatment unit.
- *E. coli* was detected in one treatment device sample
- HPC bacterial levels were four times higher in the device samples compared to the household tap samples.

RECOMMENDATIONS

- The level of *Giardia* disinfection (CxT credit) should be evaluated on a regular basis to avoid excessive levels of disinfection that could result in unnecessary increases in disinfection by product levels.
- Additional studies of coliphage occurrence and survival in drinking water should be conducted to better assess the potential for using these organisms as indicators of distribution system contamination. Research using GIS-linked mapping is necessary to establish relationships between distribution system main breaks and water quality parameters.
- Iowa-American should consider the installation of surge tanks within its next engineering review to minimize the potential for negative pressure transients within the distribution system.
- Additional monitoring using electronic pressure recorders will better assess the potential for negative pressure transients within the Davenport system. Special

emphasis should be placed on the locations where the model predicts low pressures are possible. The recorders should be set to record and send an alarm to the control center when low pressures are encountered.

- Distribution system operators should be aware that main breaks and the rapid opening of certain fire hydrants could result in low distribution system pressures. Operator training and appropriate standard operating procedures should be developed to respond to such situations.
- When the treatment device codes are released, a detailed analysis of microbial concentrations and occurrence rates for total coliform and HPC bacteria should be performed for the active and inactive (placebo) devices. Researchers should determine if there were any differences between the units that could account for microbial disease.
- Researchers should examine the illness rates of the study participants for the months just prior to, and after, the treatment device crossover. The total coliform and HPC bacterial levels should be linked to the respective households to determine if households with high microbial levels had different levels of gastrointestinal disease than families with uncontaminated units. Researchers should examine any changes in illness rates following crossover.

APPENDIX A – Julian Calendar

Table A.1 – Julian calendar for the year of 2000 (November to December), 2001 and 2002

Day	Year 2000[1]		Year 2001											
	November	December	January	February	March	April	May	June	July	August	September	October	November	December
1	30600	33600	00101	03201	06001	09101	12101	15201	18201	21301	24401	27401	30501	33501
2	30700	33700	00201	03301	06101	09201	12201	15301	18301	21401	24501	27501	30601	33601
3	30800	33800	00301	03401	06201	09301	12301	15401	18401	21501	24601	27601	30701	33701
4	30900	33900	00401	03501	06301	09401	12401	15501	18501	21601	24701	27701	30801	33801
5	31000	34000	00501	03601	06401	09501	12501	15601	18601	21701	24801	27801	30901	33901
6	31100	34100	00601	03701	06501	09601	12601	15701	18701	21801	24901	27901	31001	34001
7	31200	34200	00701	03801	06601	09701	12701	15801	18801	21901	25001	28001	31101	34101
8	31300	34300	00801	03901	06701	09801	13801	15901	18901	22001	25101	28101	31201	34201
9	31400	34400	00901	04001	06801	09901	12901	16001	19001	22101	25201	28201	31301	34301
10	31500	34500	01001	04101	06901	10001	13001	16101	19101	22201	25301	28301	31401	34401
11	31600	34600	01101	04201	07001	10101	13101	16201	19201	22301	25401	28401	31501	34501
12	31700	34700	01201	04301	07101	10201	13201	16301	19301	22401	25501	28501	31601	34601
13	31800	34800	01301	04401	07201	10301	13301	16401	19401	22501	25601	28601	31701	34701
14	31900	34900	01401	04501	07301	10401	13401	16501	19501	22601	25701	28701	31801	34801
15	32000	35000	01501	04601	07401	10501	13501	16601	19601	22701	25801	28801	31901	34901
16	32100	35100	01601	04701	07501	10601	13601	16701	19701	22801	25901	28901	32001	35001
17	32200	35200	01701	04801	07601	10701	13701	16801	19801	22901	26001	29001	32101	35101
18	32300	35300	01801	04901	07701	10801	13801	16901	19901	23001	26101	29101	32201	35201
19	32400	35400	01901	05001	07801	10901	13901	17001	20001	23101	26201	29201	32301	35301
20	32500	35500	02001	05101	07901	11001	14001	17101	20101	23201	26301	29301	32401	35401
21	32600	35600	02101	05201	08001	11101	14101	17201	20201	23301	26401	29401	32501	35501
22	32700	35700	02201	05301	08101	11201	14201	17301	20301	23401	26501	29501	32601	35601
23	32800	35800	02301	05401	08201	11301	14301	17401	20401	23501	26601	29601	32701	35701
24	32900	35900	02401	05501	08301	11401	14401	17501	20501	23601	26701	29701	32801	35801
25	33000	36000	02501	05601	08401	11501	14501	17601	20601	23701	26801	29801	32901	35901
26	33100	36100	02601	05701	08501	11601	14601	17701	20701	23801	26901	29901	33001	36001
27	33200	36200	02701	05801	08601	11701	14701	17801	20801	23901	27001	30001	33101	36101
28	33300	36300	02801	05901	08701	11801	14801	17901	20901	24001	27101	30101	33201	36201
29	33400	36400	02901		08801	11901	14901	18001	21001	24101	27201	30201	33301	36301
30	33500	36500	03001		08901	12001	15001	18101	21101	24201	27301	30301	33401	36401
31		36600	03101		09001		15101		21201	24301		30401		36501

[1] Leap year.

Table A.1, continued –Julian calendar for the year of 2000 (November to December), 2001 and 2002

Day	Year 2002											
	January	February	March	April	May	June	July	August	September	October	November	December
1	00102	03202	06002	09102	12102	15202	18202	21302	24402	27402	30502	33502
2	00202	03302	06102	09202	12202	15302	18302	21402	24502	27502	30602	33602
3	00302	03402	06202	09302	12302	15402	18402	21502	24602	27602	30702	33702
4	00402	03502	06302	09402	12402	15502	18502	21602	24702	27702	30802	33802
5	00502	03602	06402	09502	12502	15602	18602	21702	24802	27802	30902	33902
6	00602	03702	06502	09602	12602	15702	18702	21802	24902	27902	31002	34002
7	00702	03802	06602	09702	12702	15802	18802	21902	25002	28002	31102	34102
8	00802	03902	06702	09802	13802	15902	18902	22002	25102	28102	31202	34202
9	00902	04002	06802	09902	12902	16002	19002	22102	25202	28202	31302	34302
10	01002	04102	06902	10002	13002	16102	19101	22202	25302	28302	31402	34402
11	01102	04202	07002	10102	13102	16202	19202	22302	25402	28402	31502	34502
12	01202	04302	07102	10202	13202	16302	19302	22402	25502	28502	31602	34602
13	01302	04402	07202	10302	13302	16402	19402	22502	25602	28602	31702	34702
14	01402	04502	07302	10402	13402	16502	19502	22602	25702	28702	31802	34802
15	01502	04602	07402	10502	13502	16602	19602	22702	25802	28802	31902	34902
16	01602	04702	07502	10602	13602	16702	19702	22802	25902	28902	32002	35002
17	01702	04802	07602	10702	13702	16802	19802	22902	26002	29002	32102	35102
18	01802	04902	07702	10802	13802	16902	19902	23002	26102	29102	32202	35202
19	01902	05002	07802	10902	13902	17002	20002	23102	26202	29202	32302	35302
20	02002	05102	07902	11002	14002	17102	20102	23202	26302	29302	32402	35402
21	02102	05202	08002	11102	14102	17202	20202	23302	26402	29402	32502	35502
22	02202	05302	08102	11202	14202	17302	20301	23402	26502	29502	32602	35602
23	02302	05402	08202	11302	14302	17402	20402	23502	26602	29602	32702	35702
24	02402	05502	08302	11402	14402	17502	20502	23602	26702	29702	32802	35802
25	02502	05602	08402	11502	14502	17602	20602	23702	26802	29802	32902	35902
26	02602	05702	08502	11602	14602	17702	20702	23802	26902	29902	33002	36002
27	02702	05802	08602	11702	14702	17802	20802	23902	27002	30002	33102	36102
28	02802	05902	08702	11802	14802	17902	20902	24002	27102	30102	33202	36202
29	02902		08802	11902	14902	18002	21002	24102	27202	30202	33302	36302
30	03002		08902	12002	15002	18102	21102	24202	27302	30302	33402	36402
31	03102		09002		15102		21202	24302		30402		36502

APPENDIX B – Sample Schedule and Transfer Forms

Table B.1 Sample Collection and Monitoring Frequency

Task	Description/ /Method	Sampling frequency/ Total number of samples	Laboratory/ Personnel
Raw Water Site Characterization	Site characterization has already taken place under the context of AWWARF #488.		AW
Raw Water Monitoring			
Raw water sampling	Grab sample/Filtration	Weekly/monthly	Iowa-Am/ Jim Case
Total and fecal coliforms and *E. coli*	Standard Methods	Weekly/ (total70 sample)	Iowa-Am
Bacillus	Spread plate method/MF	Weekly/(total 70 samples)	AW
Clostridium	Membrane Filtration	Weekly/(total 70 samples)	AW
Coliphage (using host *E. coli* CN-13	Membrane Filtration	Weekly/(total 70 samples)	AW
Coliphage (using host *E. coli* F-Amp	Membrane Filtration	Weekly/(total 70 samples)	AW
Cultural viruses	ICR methodology	Monthly (total 16 samples)	AW
Giardia	Method 1623	[1]Weekly/(total 70 samples.	CEC
Cryptosporidium	Method 1623 and CC-PCR	[1]Weekly	CEC and AW
Algae		[1]Weekly	CEC

[1] A single sample will be analyzed for *Giardia, Cryptosporidium*, and algae (continued)

Table B.1 Sample Collection and Monitoring Frequency (cont.)

Task	**Description/ /Method**	**Sampling frequency/ Total samples**	**Laboratory/ Personnel**
Treatment Plant Characterization	Compilation of information from plant description book, engineering records, and utility data.		AW
Treatment Plant Monitoring			
Electronic Purification Reports	Daily purification log	Hourly, summarized daily	Iowa-Am
Filter Operation Log	For each individual filter, continuous monitoring with recording of hourly values, used to calculate a daily average.	Continuous, hourly, summarized daily	Iowa-Am
Form 100	Plant Operational parameters	summarized daily	
Raw water sampling	Grab sample/Filtration	Weekly/monthly	Iowa-Am
Total coliforms	Standard Methods	Daily	Iowa-Am
HPC[2]	Spread plate method	Daily	Iowa-Am
Coliphage (using host E. coli CN-13	Membrane Filtration	Weekly	Iowa-Am
Coliphage (using host E. coli F-Amp	Membrane Filtration	Weekly	Iowa-Am
Bacillus	Spread plate method/MF	Weekly/(total 70 samples)	Iowa-Am
Clostridium	Membrane Filtration	Weekly/(total 70 samples)	AW
Cultural viruses	ICR methodology	Monthly (total 16 samples)	AW
Cryptosporidium	CC-PCR	Weekly/(total 70 samples)	AW
			AW

[2] Media will be prepared at the Belleville lab and shipped to Iowa-Am

(continued)

Table B.1 Sample Collection and Monitoring Frequency (cont.)

Task	Description/ /Method	Sampling frequency/ Total number of samples	Laboratory/ Personnel
Distribution System Characterization	Distribution system description book (Maps, diagrams etc.)		AW
Hydrogeological Model	Modeling software (pipe 2000, Surge 2000)		U of Kentucky
Distribution system leaks and repairs	Iowa-Am log of leaks and repairs		Iowa-Am
Distribution System Monitoring			
Routine monitoring (Total and fecal coliforms, HPC[2], chlorine, pH, turbidity, phosphate, and temperature)	Standard or acceptable methods	Daily (110-120 samples per month)	Iowa-Am
Coliphage (using host *E. coli* CN-13	Membrane Filtration	Weekly	AW
Coliphage (using host *E. coli* F-Amp	Membrane Filtration	Weekly	AW
Transient pressure monitoring	Radcom pressure data logger	Continuous for one year	AW/ Iowa-Am
Hydraulic modeling for surge vulnerability	Surge software		U of Kentucky

[2]Media will be prepared at the Belleville lab and shipped to Iowa-Am (continued)

Table B.1 Sample Collection and Monitoring Frequency (cont.)

Task	Description/Method	Sampling frequency/ Total number of samples	Laboratory/ Personnel
Monitoring Household Tapes and Treatment Devices			
Sample collection	Samples from both kitchen faucet tap water and device water	Three times during study (total 600 samples	Epidemiology team
Total chlorine, pH, temperature,	Acceptable methods	On site	Epidemiology team
Total coliforms, HPC[2], turbidity	Standard/acceptable methods	Three time during the study (total 600 samples)	Iowa-Am
Data Management			TPMC
Data Analysis			TPMC
Data Transfer			TPMC
Report preparation			AW

AWWARF 2580 – Infectious Disease Associated with Drinking Water from Surface Water Sources – Microbiological Water Quality Factors

SAMPLE COLLECTION SHEET FOR COLIPHAGE, *CLOSTRIDIUM PERFRINGENS*, CULTURABLE VIRUSES, AND *BACILLUS SUBTILUS*

Utility: ______________________________

Sampler: ______________________ Date Sampled: __________________

Utility Contact: ____

Telephone No: ____________ **Fax:** ______________________________

E-mail address: _______________________

Sampling Data

Sample Collection Time: ________________________________

Comments: List any other information about the sample collection event that may be important.

____
____
____
____

Please send samples to:

American Water Works Service Co.
1115 S. Illinois St.
Belleville, IL 62220
Attn: Auth. 219
(618) 235-3600

Sample Code

Coliphage ______________________________________

Clostridium perfringens ______________________________

Bacillus subtilis ______________________________________

Culturable viruses ______________________________________

Figure B.1 – Sample transfer form received by AW from East River Station, Davenport, IA

AWWARF 2580 – Infectious Disease Associated with Drinking Water from Surface Water Sources – Microbiological Water Quality Factors

SAMPLE COLLECTION SHEET FOR *GIARDIA/CRYPTOSPORIDIUM*

Utility: ____________________________

Sampler: ____________________________ Date Sampled: _______________

Utility Contact: ____

Telephone No: ______________ **Fax:** ____________________________

E-mail address: ____________________________

Sampling Data

Start Time: ________________________ End Time: ____________________________

Meter Reading start/end ____

Comments: List any other information about the sample collection event that may be important.

____
____
____
____
____
____

Please send samples to: **Clancy Environmental Consultants, Inc.**
272 North Main St.
St. Albans, VT 05478
Phone: (802) 527-2460

Sample Code ____

Figure B.2 - Sample transfer form received by CEC from East River Station, Davenport, IA.

AWWARF 2580 – Infectious Disease Associated with Drinking Water from Surface Water Sources – Microbiological Water Quality Factors

SAMPLE TRANSFER SHEET FROM CEC TO AW FOR *GIARDIA/CRYPTOSPORIDIUM*

Sample Code ______________________

Utility ______________________________

Date Sampled_______________________________

Date Shipped to AW__________________________

Total Packed Pellet Volume (mL)__________________________

Packed Pellet Volume (mL) Sent to AW_______________________

Packed Pellet Volume Equivalent (L) Sent to AW_________________

Comments: List any other information about the sample that may be important.

____
____
____
____
____

Figure B.3 - Sample transfer form between CEC and AW

APPENDIX C - Laboratory Bench Sheets

AWWARF 2580 – Infectious Disease Associated with Drinking Water from Surface Water Sources – Microbiological Water Quality Factors

ALGAE LABORATORY BENCHSHEET

Sample ID	Utility
PPV analyzed: μL	Date analyzed/initials:
Equivalent volume: L	

MICROSCOPIC EXAMINATION

Algae ID	Count/field	Count/slide	Total Count/L

Figure C.1 - Laboratory bench sheet for algae analysis performed at CEC.

AWWARF 2580 – Infectious Disease Associated with Drinking Water from Surface Water Sources – Microbiological Water Quality Factors

CRYPTOSPORIDIUM/GIARDIA LABORATORY BENCHSHEET

Sample ID	Utility
Date Received/Initials	Sample Date
Filter Lot #	Volume filtered L
Comments	
SAMPLE PROCESSING	
Elution Analyst /Date:	PPV: mL
PPV assayed: mL	Equivalent volume assayed: L
Analyst/Date:	Dynal Lot #
MICROSCOPIC EXAMINATION	
Well Slide Lot #	G/C mAb Lot#
Analyst/Date:	Microscope:
DAPI Lot #	

Object	Size (μm)	DAPI Observations	DIC Observations

Figure C.2 - Laboratory bench sheet for *Giardia* and *Cryptosporidium* analyses performed at CEC.

APPENDIX D – Data Collection Sheets

AWWARF 2580 Data Collection Sheet - Protozoan and Algae	
Sample ID	
Sample date	
Date received CEC	
Sample volume (L)	
Elution date	
Packed pellet volume (mL)	
Pellet volume processed by IMS (mL)	
IMS date	
IMS concentrate volume (μL)	
IMS concentrate volume sent to AW (μL)	
Equivalent sample volume sent to AW (L)	
Date shipped to AW	
IMS concentrate volume analyzed at CEC (μL)	
Equivalent sample volume by IMS (L)	
Date stained	
Date analyzed	
Total *Giardia* counted	
Total *Giardia* per L	
Observed characteristics:	
Total *Giardia* with axonemes counted	
Total *Giardia* with median bodies counted	
Total *Giardia* with nuclei counted	
Total *Cryptosporidium* counted	
Total *Cryptosporidium* per L	
Observed characteristics:	
Total *Cryptosporidium* with sporozoites counted	
Total empty *Cryptosporidium* counted	
Total pre-IMS volume (mL)	
Actual volume analyzed for algae (μL)	
Equivalent volume analyzed (L)	
Number fields counted	
Total number fields	
Algae count	
Total algae count	
Total algae per L	

Figure D.1 – Data collection sheet for protozoan and algae sample analyzed by CEC.

AWWARF 2580 Data Collection Sheet - *Bacillus subtilis*	
Sample ID	
Date of Collection	
Date Received	
Date Filtered/Plated	
Amount Plated Out (mL)	
Volume Filtered (mL)	
Bacillus Mean CFU	
Bacillus Mean CFU/100 mL	
Analyst Initials	
Bacillus Qualifier	

Figure D.2 – Data collection sheet for *Bacillus subtilis* sample analyzed by AW

.

AWWARF 2580 Data Collection Sheet - CC-PCR	
Sample ID	
Date of Collection	
Date Received	
Date Dissociated	
Analyst Initials	
CC-PCR Date	
CC-PCR Result	
Analyst Initials	
CC-PCR Qualifier	

Figure D.3 – Data collection sheet for CC-PCR sample analyzed by AW.

AWWARF 2580 Data Collection Sheet - *Clostridium perfringens*	
Sample ID	
Date of Collection	
Date Received	
Date Filtered	
Volume filtered (mL)	
Clostridium CFU	
Clostridium CFU/100 mL	
Analyst Initials	
Clostridium Qualifier	

Figure D.4 – Data collection sheet for *Clostridium perfringens* sample analyzed by AW.

AWWARF 2580 Data Collection Sheet - Coliphage Sample	
Sample ID	
Date of Collection	
Date Received	
Date Sample Assayed by SAL Method	
Total Volume Assayed (mL)	
Mean Male Specific Coliphage (PFU/100 mL)	
Mean Somatic Coliphage (PFU/100 mL)	
Analyst Initials	
Phage Qualifier	

Figure D.5 – Data collection sheet for coliphage sample analyzed by AW

AWWARF 2580 Data Collection Sheet – Culturable Viruses Sample	
Sample ID	
Date of Collection	
Date Received	
Date Eluted	
Date Concentrated	
Volume of Water Filtered (L)	
Final Volume of Sample Concentrate (mL)	
Assay Sample Volume (mL)	
Volume of Original Water Sample Assayed (L)	
Date Assayed	
Virus flag	
MPN/100 L	
Analyst Initials	
Virus Qualifier	

Figure D.6 – Data collection sheet for culturable viruses sample analyzed by AW

AWWARF 2580 Data Collection Sheet - Distribution System	
Sample ID	
Date of Collection	
Total flag	
Total coliform (cfu/100mL)	
HPC flag	
HPC bacteria (cfu/mL)	
Total chlorine (mg/L)	
Free chlorine (mg/L)	
pH (pH units)	
Turbidity (ntu)	
Phosphate (mg/L)	
Temperature (°C)	

Figure D.7 – Data collection sheet for the distribution system at the East River Station, Davenport, IA

AWWARF 2580 Data Collection Sheet - Households	
Sample ID	
Date of Collection	
Total flag	
Total coliform (cfu/100mL)	
HPC flag	
HPC bacteria (cfu/mL)	
Total chlorine (mg/L)	
pH (pH units)	
Turbidity (ntu)	
Temperature (°C)	

Figure D.8 – Data collection sheet for the households, Davenport, IA

AWWARF 2580 Data Collection Sheet – Plant Operation, daily average values	
Sample ID	
Date of collection	
Plant Flow (mgd)	
Raw	
Superpulsator	
Sedimentation Tank 1	
Sedimentation Tank 2	
Filter 1	
Filter 2	
Filter 3	
Filter 4	
Filter 5	
Filter 6	
Filter 7	
Filter 8	
Filter 9	
Filter 10	
Filter 11	
Filter 12	
Filter 13	
Filter 14	
Filter 15	
Filter 16	
Filter 17	
Filter 18	
Filter 19	
Filter 20	
Headloss (ft)	
Filter 1	
Filter 2	
Filter 3	
Filter 4	
Filter 5	
Filter 6	
Filter 7	
Filter 8	
Filter 9	
Filter 10	
Filter 11	
Filter 12	
Filter 13	
Filter 14	
Filter 15	

Figure D.9 – Data collection sheet for East River Station Operations, Davenport, IA

AWWARF 2580 Data Collection Sheet – Plant Operation, continuation	
Filter 16	
Filter 17	
Filter 18	
Filter 19	
Filter 20	
Number of filters washed (units)	
Average filter run (hours)	
Wash water (mgd)	
Fluoride liq. HFS acid (lbs/mgd)	
Fluoride dry HFS acid (lbs/mgd)	
Chlorine gas liq. (lbs/mgd)	
Ferric Cl wet (lbs/mgd)	
PAS A410 wet (lbs/mgd)	
Ammonia gas liq. (lbs/mgd)	
Pot. Perm. dry (dry/mgd)	
Salt dry (lbs/mgd)	
Zinc PO4 liq. (lbs/mgd)	
Solids blanket polymer liq. (lbs/mgd)	
Filter aid polymer liq. (lbs/mgd)	
Caustic soda wet (lbs/mgd)	
Caustic soda dry (lbs/mgd)	
Cationic polymer (lbs/mgd)	

Figure D.9, continuation – Data collection sheet for East River Station Operations, Davenport, IA

AWWARF 2580 Data Collection Sheet – Effluent	
Sample ID	
Date of collection	
Particle count (#/mL)	
Turbidity (ntu)	
Alkalinity (mg/L as $CaCO_3$)	
Hardness (mg/L as $CaCO_3$)	
pH (pH units)	
Fluoride (mg/L)	
Phosphate (mg/L)	
Total chlorine (mg/L)	
Total flag	
Total coliform (cfu/100 mL)	
HPC flag	
HPC bacteria (cfu/mL)	
Flow (mgd)	
CT obtained/CT required	

Figure D.10 – Data collection sheet for the effluent at the East River Station, Davenport, IA

AWWARF 2580 Data Collection Sheet – Raw water	
Sample ID	
Date of collection	
pH (pH units)	
Temperature (°C)	
Turbidity (ntu)	
Alkalinity (mg/L as $CaCO_3$)	
Hardness (mg/L as $CaCO_3$)	
Fecal flag	
Fecal coliform (cfu/100 mL)	
Total flag	
Total coliform (cfu/100 mL)	
River flow (cfs)	
River stage (ft)	
Precipitation (inches)	

Figure D.11 – Data collection sheet for the raw water at the East River Station, Davenport, IA

AWWARF 2580 Data Collection Sheet – Treatment Plant	
Sample ID	
Date of collection	
Particle count – Superpulsator (#/mL)	
Turbidity (ntu)	
Superpulsator 1	
Superpulsator 2	
Average Superpulsator 1 & 2	
Sedimentation Tank 1	
Sedimentation Tank 2	
Sedimentation Tank 1 and 2 and Superpulsator 1 & 2	
Filter 1	
Filter 2	
Filter 3	
Filter 4	
Filter 5	
Filter 6	
Filter 7	
Filter 8	
Filter 9	
Filter 10	
Filter 11	
Filter 12	
Filter 13	
Filter 14	
Filter 15	
Filter 16	
Filter 17	
Filter 18	
Filter 19	
Filter 20	
Alkalinity (mg/L as $CaC0_3$)	
Sedimentation Tank 1	
Sedimentation Tank 2	
pH (pH units)	
Sedimentation Tank 1	
Sedimentation Tank 2	
Superpulsator 1	
Superpulsator 2	
Fluoride (mg/L) – Dist. Tap	
Phosphate (mg/L) – Dist. Tap	
Free chlorine (mg/L)	
Superpulsator	
Sedimentation Tank 1	

Figure D.12 – Data collection sheet for East River Station, Davenport, IA

AWWARF 2580 Data Collection Sheet – Treatment Plant, continuation	
Sedimentation Tank 2	
Average Superpulsator and Sedimentation Tank 1 & 2	
Total chlorine (mg/L) – Dist. Tap	

Figure D.12, continuation – Data collection sheet for East River Station, Davenport, IA

APPENDIX E – Database Design

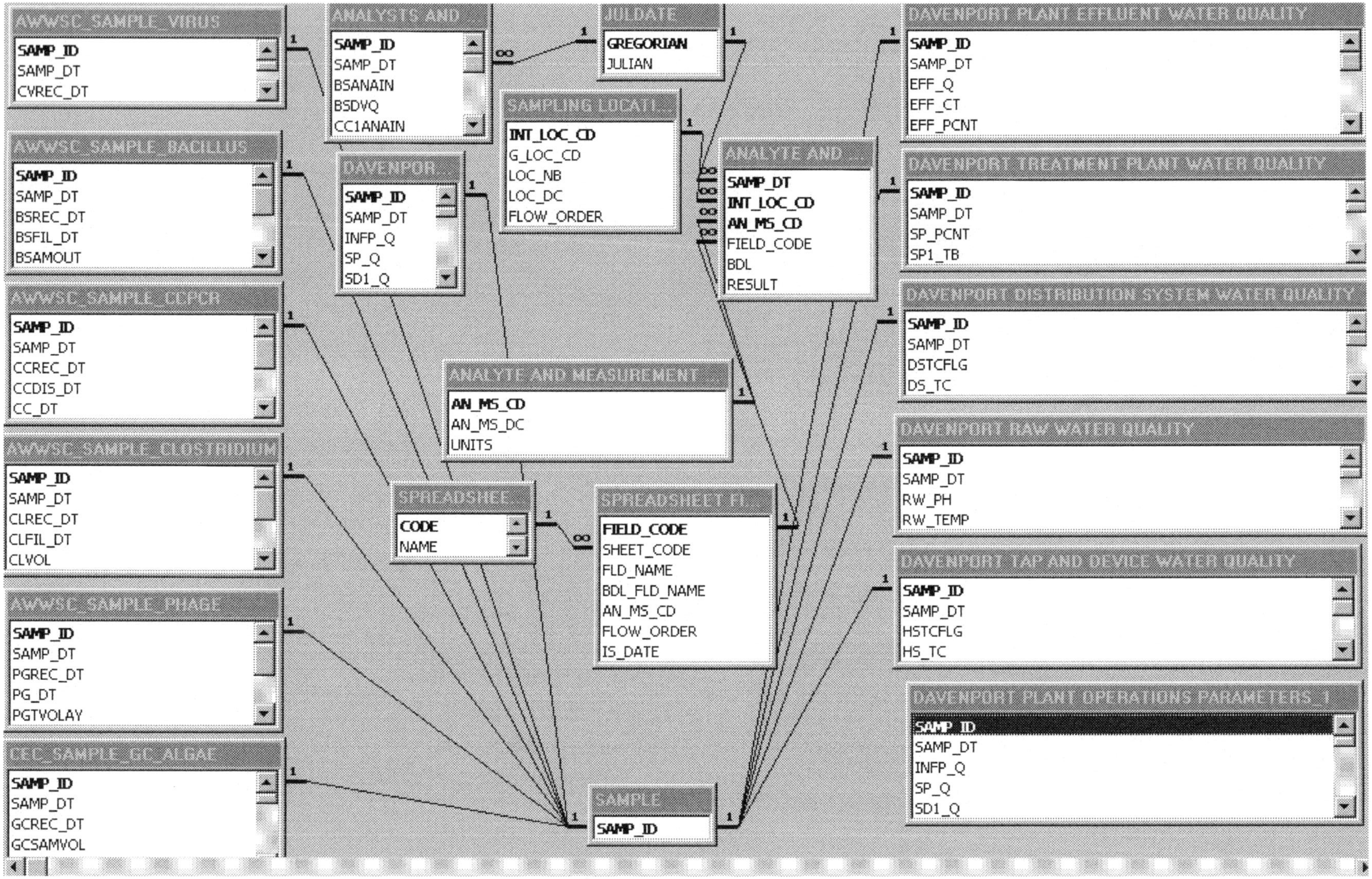

Figure E.1 – Entity Relational Diagram (ERD) for AwwaRF Project 2580 Database

Table Name: DAVENPORT RAW WATER QUALITY

Attribute	Units	Format	Minimum	Maximum	Description
SAMP_ID	n/a	C (11)	n/a	n/a	Sample ID
SAMP_DT	n/a	mm/dd/yyyy	n/a	n/a	Sample Date
RW_PH	pH units	N (4.2)	0.00	14.00	Raw Water pH
RW_TEMP	°C	N (4.2)	0.00	35.00	Raw Water Temperature
RW_TB	NTU	N (6.2)	0.00	2000.00	Raw Water Turbidity
RW_ALK	mg/L as $CaCO_3$	N (4.1)	0.0	300.0	Raw Water Alkalinity
RW_HARD	mg/L as $CaCO_3$	N (4.1)	0.0	300.0	Raw Water Hardness
RWFCFLG	n/a	C (1)	n/a	n/a	Raw Water Fecal Flag
RW_FC	CFU/ 100 mL	N (5)	0	1×10^4	Raw Water Fecal Coliform
RWTCFLG	n/a	C (1)	n/a	n/a	Raw Water Total Flag
RW_TC	CFU/ 100 mL	N (6)	0	1×10^5	Raw Water Total Coliform
RIVER_Q	cubic feet/second	N (7)	0	1×10^6	River Flow (cfs)
RIVER_ST	feet	N (4.2)	0	10.00	River Stage
PRECIP	inches	N (3.2)	0.00	5.00	Precipitation

Figure E.2 – Davenport Raw Water Quality Entity

Table Name: DAVENPORT TREATMENT PLANT WATER QUALITY

Attribute	Units	Format	Minimum	Maximum	Description
SAMP_ID	n/a	C (11)	n/a	n/a	Sample ID
SAMP_DT	n/a	mm/dd/yyyy	n/a	n/a	Sample Date
SP_PCNT	Particles/mL	N (6.2)	0.00	5000.00	SP Particle Count
SP1_TB	NTU	N (3.2)	0.00	3.00	SP # 1 Turbidity
SP2_TB	NTU	N (3.2)	0.00	3.00	SP # 2 Turbidity
SD1_TB	NTU	N (3.2)	0.00	3.00	SED # 1 Turbidity
SD2_TB	NTU	N (3.2)	0.00	3.00	SED # 2 Turbidity
FL1_TB	NTU	N (3.2)	0.00	3.00	Filter # 1 Turbidity
FL2_TB	NTU	N (3.2)	0.00	3.00	Filter # 2 Turbidity
FL3_TB	NTU	N (3.2)	0.00	3.00	Filter # 3 Turbidity
FL4_TB	NTU	N (3.2)	0.00	3.00	Filter # 4 Turbidity
FL5_TB	NTU	N (3.2)	0.00	3.00	Filter # 5 Turbidity
FL6_TB	NTU	N (3.2)	0.00	3.00	Filter # 6 Turbidity
FL7_TB	NTU	N (3.2)	0.00	3.00	Filter # 7 Turbidity
FL8_TB	NTU	N (3.2)	0.00	3.00	Filter # 8 Turbidity
FL9_TB	NTU	N (3.2)	0.00	3.00	Filter # 9 Turbidity
FL10_TB	NTU	N (3.2)	0.00	3.00	Filter # 10 Turbidity
FL11_TB	NTU	N (3.2)	0.00	3.00	Filter # 11 Turbidity
FL12_TB	NTU	N (3.2)	0.00	3.00	Filter # 12 Turbidity
FL13_TB	NTU	N (3.2)	0.00	3.00	Filter # 13 Turbidity
FL14_TB	NTU	N (3.2)	0.00	3.00	Filter # 14 Turbidity
FL15_TB	NTU	N (3.2)	0.00	3.00	Filter # 15 Turbidity
FL16_TB	NTU	N (3.2)	0.00	3.00	Filter # 16 Turbidity
FL17_TB	NTU	N (3.2)	0.00	3.00	Filter # 17 Turbidity
FL18_TB	NTU	N (3.2)	0.00	3.00	Filter # 18 Turbidity
FL19_TB	NTU	N (3.2)	0.00	3.00	Filter # 19 Turbidity
FL20_TB	NTU	N (3.2)	0.00	3.00	Filter # 20 Turbidity

Figure E.3 – **Davenport Treatment Plant Water Quality Entity**

Table Name, cont'd: DAVENPORT TREATMENT PLANT WATER QUALITY

Attribute	Units	Format	Minimum	Maximum	Description
SD1_ALK	mg/L as $CaCO_3$	N (4.1)	0.0	300.0	SED. # 1 Alkalinity
SD2_ALK	mg/L as $CaCO_3$	N (4.1)	0.0	300.0	SED. # 2 Alkalinity
SP1_PH	(pH units)	N (4.2)	0.00	14.00	SP # 1 pH
SP2_PH	(pH units)	N (4.2)	0.00	14.00	SP # 2 pH
SD1_PH	(pH units)	N (4.2)	0.00	14.00	SED. # 1 pH
SD2_PH	(pH units)	N (4.2)	0.00	14.00	SED. # 2 pH
DTP_FL	(pH units)	N (3.2)	0.00	5.00	Fluoride Distant Tap
DTP_PO4	(mg/L)	N (3.2)	0.00	5.00	Phosphate Distant Tap
SP_FCL	(mg/L)	N (3.2)	0.00	5.00	Free Chlorine SP
SD1_FCL	(mg/L)	N (3.2)	0.00	5.00	Free Chlorine SED. # 1
SD2_FCL	(mg/L)	N (3.2)	0.00	5.00	Free Chlorine SED. # 2
SPSDGFCL	(mg/L)	N (3.2)	0.00	5.00	Free Chlorine SP and SED. AVG.
DTP_TCL	(mg/L)	N (3.2)	0.00	5.00	Total Chlorine Distant Tap
EFF_Q	MGD	N (3.2)	0.00	3.00	Finished Flow

Figure E.3 – Davenport Treatment Plant Water Quality Entity

Table Name: DAVENPORT PLANT EFFLUENT WATER QUALITY

Attribute	Units	Format	Minimum	Maximum	Description
SAMP_ID	n/a	C (11)	n/a	n/a	Sample ID
SAMP_DT	n/a	mm/dd/yyyy	n/a	n/a	Sample Date
EFF_PCNT	Particles/mL	N (4.2)	0.00	50.00	Effluent Particle Count
EFF_TB	(NTU)	N (3.2)	0.00	3.00	Effluent Turbidity
EFF_ALK	mg/L as $CaCO_3$	N (4.1)	0.00	300.0	Effluent Alkalinity
EFF_HARD	mg/L as $CaCO_3$	N (4.1)	0.0	300.0	Effluent Hardness
EFF_pH	pH units	N (4.2)	0.00	14.00	Effluent pH
EFF_FL	mg/L	N (3.2)	0.00	5.00	Effluent Fluoride
EFF_PO4	mg/L	N (3.2)	0.00	5.00	Effluent Phosphate
EFF_TCL	mg/L	N (3.2)	0.00	5.00	Effluent Total Chlorine
EFFTCFLG	n/a	C (1)	n/a	n/a	Effluent Total Flag
EFF_TC	CFU/100 mL	N (3)	0	100	Effluent Total Coliform
EFFHPFLG	n/a	C (1)	n/a	n/a	Effluent HPC Flag
EFF_HPC	CFU/mL	N (4)	0	2000	Effluent HPC
EFF_Q	MGD	N (3.2)	0.00	3.00	Finished Flow
EFF_CT	n/a	N (4.2)	0.00	10.00	Plant Effluent CT Obtained/Ct Required

Figure E.4 – Davenport Plant Effluent Water Quality Entity

Table Name: DAVENPORT PLANT OPERATIONS PARAMETERS

Attribute	Units	Format	Minimum	Maximum	Description
SAMP_ID	n/a	C (11)	n/a	n/a	Sample ID
SAMP_DT	n/a	mm/dd/yyyy	n/a	n/a	Sample Date
INF_Q	MGD	N (4.2)	0.00	40.00	Influent Plant Flow
SP_Q	MGD	N (4.2)	0.00	40.00	SP Flow
SD1_Q	MGD	N (4.2)	0.00	40.00	SED. # 1 Flow
SD2_Q	MGD	N (4.2)	0.00	40.00	SED. # 2 Flow
FL1_Q	MGD	N (3.2)	0.00	3.00	Filter # 1 Flow
FL2_Q	MGD	N (3.2)	0.00	3.00	Filter # 2 Flow
FL3_Q	MGD	N (3.2)	0.00	3.00	Filter # 3 Flow
FL4_Q	MGD	N (3.2)	0.00	3.00	Filter # 4 Flow
FL5_Q	MGD	N (3.2)	0.00	3.00	Filter # 5 Flow
FL6_Q	MGD	N (3.2)	0.00	3.00	Filter # 6 Flow
FL7_Q	MGD	N (3.2)	0.00	3.00	Filter # 7 Flow
FL8_Q	MGD	N (3.2)	0.00	3.00	Filter # 8 Flow
FL9_Q	MGD	N (3.2)	0.00	3.00	Filter # 9 Flow
FL10_Q	MGD	N (3.2)	0.00	3.00	Filter # 10 Flow
FL11_Q	MGD	N (3.2)	0.00	3.00	Filter # 11 Flow
FL12_Q	MGD	N (3.2)	0.00	3.00	Filter # 12 Flow
FL13_Q	MGD	N (3.2)	0.00	3.00	Filter # 13 Flow
FL14_Q	MGD	N (3.2)	0.00	3.00	Filter # 14 Flow
FL15_Q	MGD	N (3.2)	0.00	3.00	Filter # 15 Flow
FL16_Q	MGD	N (3.2)	0.00	3.00	Filter # 16 Flow
FL17_Q	MGD	N (3.2)	0.00	3.00	Filter # 17 Flow
FL18_Q	MGD	N (3.2)	0.00	3.00	Filter # 18 Flow
FL19_Q	MGD	N (3.2)	0.00	3.00	Filter # 19 Flow
FL20_Q	MGD	N (3.2)	0.00	3.00	Filter # 20 Flow
FL1_HL	FT	N (3.2)	0.00	3.00	Filter # 1 Headloss
FL2_HL	FT	N (3.2)	0.00	3.00	Filter # 2 Headloss
FL3_HL	FT	N (3.2)	0.00	3.00	Filter # 3 Headloss
FL4_HL	FT	N (3.2)	0.00	3.00	Filter # 4 Headloss
FL5_HL	FT	N (3.2)	0.00	3.00	Filter # 5 Headloss
FL6_HL	FT	N (3.2)	0.00	3.00	Filter # 6 Headloss
FL7_HL	FT	N (3.2)	0.00	3.00	Filter # 7 Headloss
FL8_HL	FT	N (3.2)	0.00	3.00	Filter # 8 Headloss
FL9_HL	FT	N (3.2)	0.00	3.00	Filter # 9 Headloss
FL10_HL	FT	N (3.2)	0.00	3.00	Filter # 10 Headloss
FL11_HL	FT	N (3.2)	0.00	3.00	Filter # 11 Headloss
FL12_HL	FT	N (3.2)	0.00	3.00	Filter # 12 Headloss
FL13_HL	FT	N (3.2)	0.00	3.00	Filter # 13 Headloss
FL14_HL	FT	N (3.2)	0.00	3.00	Filter # 14 Headloss
FL15_HL	FT	N (3.2)	0.00	3.00	Filter # 15 Headloss
FL16_HL	FT	N (3.2)	0.00	3.00	Filter # 16 Headloss

Figure E.5 – Davenport Plant Operations Parameters Entity

Table Name, cont'd: DAVENPORT PLANT OPERATIONS PARAMETERS

Attribute	Units	Format	Minimum	Maximum	Description
FL17_HL	FT	N (3.2)	0.00	3.00	Filter # 17 Headloss
FL18_HL	FT	N (3.2)	0.00	3.00	Filter # 18 Headloss
FL19_HL	FT	N (3.2)	0.00	3.00	Filter # 19 Headloss
FL20_HL	FT	N (3.2)	0.00	3.00	Filter # 20 Headloss
NFL_WSH	n/a	N (2)	0	20	Number of Filters Washed
NFL_WSH	Hours	N (3.1)	0.0	60.0	Avg. Filter Run
WSWAT_Q	MGD	N (3.2)	0.00	2.00	Wash Water Flow
LIQFL_DS	LBS/MGD	N (5.1)	0.0	1000.0	Liquid Fluoride Dose
DRYFL_DS	LBS/MGD	N (5.1)	0.0	1000.0	Dry Fluoride Dose
GLCL_DS	LBS/MGD	N (5.1)	0.0	1000.0	Gas Liquid Chlorine Dose
WFCL_DS	LBS/MGD	N (5.1)	0.0	1000.0	Wet Ferric Cl Dose
PAS_DS	LBS/MGD	N (5.1)	0.0	1000.0	Wet PAS A410 Dose
GLNH3_DS	LBS/MGD	N (5.1)	0.0	1000.0	Gas Liquid Ammonia Dose
KMN04_DS	LBS/MGD	N (5.1)	0.0	1000.0	Dry KMnO4 Dose
SALTD_DS	LBS/MGD	N (5.1)	0.0	1000.0	Salt Dry Dose
ZPO4_DS	LBS/MGD	N (5.1)	0.0	1000.0	Liquid Zinc PO4 Dose
BLPOL_DS	LBS/MGD	N (5.1)	0.0	1000.0	Liquid Solids Blanket Polymer Dose
FLAID_DS	LBS/MGD	N (5.1)	0.0	1000.0	Liquid Filter Aid Polymer Dose
WSOD_DS	LBS/MGD	N (5.1)	0.0	1000.0	Wet Caustic Soda Dose
DSOD_DS	LBS/MGD	N (5.1)	0.0	1000.0	Dry Caustic Soda Dose
CTPOL_DS	LBS/MGD	N (5.1)	0.0	1000.0	Cationic Polymer Dose

Figure E.5 – Davenport Plant Operations Parameters Entity

Table Name: DAVENPORT DISTRIBUTION SYSTEM WATER QUALITY

Attribute	Units	Format	Minimum	Maximum	Description
SAMP_ID	n/a	C (11)	n/a	n/a	Sample ID
SAMP_DT	n/a	mm/dd/yyyy	n/a	n/a	Sample Date
DSTCFLG	n/a	C (1)	n/a	n/a	Distribution System Total Flag
DS_TC	CFU/100 mL	N (2)	0	10	Distribution System Total Coliform
DSHPFLG	n/a	C (1)	n/a	n/a	Distribution System HPC Flag
DS_HPC	CFU/ mL	N (4)	0	2000	Distribution System HPC
DS_TCL	mg/L	N (3.2)	0.00	5.00	Distribution System Total Chlorine
DS_FCL	mg/L	N (3.2)	0.00	5.00	Distribution System Free Chlorine
DS_PH	pH units	N (4.2)	0.00	14.00	Distribution System pH
DS_TB	NTU	N (3.2)	0.00	3.00	Distribution System Turbidity
DS_PO4	mg/L	N (3.2)	0.00	5.00	Distribution System Phosphate
DS_TEMP	°C	N (4.2)	0.00	35.00	Distribution System Temperature

Figure E.6 – Davenport Distribution System Water Quality Entity

Table Name: DAVENPORT TAP AND DEVICE WATER QUALITY

Attribute	Units	Format	Minimum	Maximum	Description
SAMP_ID	n/a	C (11)	n/a	n/a	Sample ID
SAMP_DT	n/a	mm/dd/yyyy	n/a	n/a	Sample Date
HSTCFLG	n/a	C (1)	n/a	n/a	Household Total Flag
HS_TC	CFU/100 mL	N (2)	0	10	Household Total Coliform
HSHPFLG	n/a	C (1)	n/a	n/a	Household HPC Flag
HS_HPC	CFU/ mL	N (4)	0	2000	Household HPC
HS_TCL	mg/L	N (3.2)	0.00	5.00	Household Total Chlorine
HS_PH	pH units	N (4.2)	0.00	14.00	Household pH
HS_TB	NTU	N (3.2)	0.00	3.00	Household Turbidity
HS_TEMP	°C	N (4.2)	0.00	35.00	Household Temperature

Figure E.7 – Davenport Tap and Device Water Quality Entity

Table Name: CEC ENVIRONMENTAL SAMPLE

Attribute	Units	Format	Minimum	Maximum	Description
SAMP_ID	n/a	C (11)	n/a	n/a	Sample ID
SAMP_DT	n/a	mm/dd/yyyy	n/a	n/a	Sample Date
GCREC_DT	n/a	mm/dd/yyyy	n/a	n/a	GC Date Received CEC
GCSAMVOL	L	N (6.3)	10.000	100.000	GC Sample Volume
GCELU_DT	n/a	mm/dd/yyyy	n/a	n/a	GC Elution Date
GCPELVOL	mL	N (5.3)	0.500	20.000	GC Packed Pellet Volume
GCIMPELV	mL	N (5.3)	0.500	20.000	GC Pellet Volume Processed by IMS
GCIM_DT	n/a	mm/dd/yyyy	n/a	n/a	GC IMS Date
GCIMVOL	μL	N (6.3)	0.050	500.000	GC IMS Concentrate Volume
IMVOLSC	μL	N(6.3)	0.050	100.000	IMS Concentrate Volume Sent to AW
EQVOLSC	L	N(6.3)	10.000	100.000	Equivalent Sample Volume Sent to AW
SHSC_DT	n/a	mm/dd/yyyy	n/a	n/a	Date Shipped to AW
GCIMVCEC	μL	N (6.3)	0.050	100.000	GC IMS Concentrate Volume Analyzed at CEC
GCIMEQV	L	N(6.3)	10.000	100.000	GC Equivalent Sample Volume by IMS
GCSTA_DT	n/a	mm/dd/yyyy	n/a	n/a	GC Date Stained
GCANA_DT	n/a	mm/dd/yyyy	n/a	n/a	GC Date Analyzed, *Giardia* and *Cryptosporidium*
GTOT	Cell number	N (2)	0	10	Total *Giardia* Counted
GTOTVOL	Cell number/L	N (5.3)	0.000	20.000	Total *Giardia* per 1 L
GDVQ	n/a	C (1)	n/a	n/a	*Giardia* Data Value Qualifier (a through z)
GAXON	Cell number	N (3)	0	10	Total *Giardia* with Axonemes Counted
GMEBOD	Cell number	N (3)	0	10	Total *Giardia* with Median Bodies Counted
GNUC	Cell number	N (2)	0	10	Total *Giardia* with Nuclei Counted
GEMPTY	Cell number	N (2)	0	10	Total Empty *Giardia* Counted
GAMORP	Cell number	N (2)	0	10	Total Amorphous *Giardia* counted
GVNTDISK	Cell number	N (2)	0	10	*Giardia* ventral disks

Table Name, cont'd: CEC ENVIRONMENTAL SAMPLE

Attribute	Units	Format	Minimum	Maximum	Description
TOTGINTS	Cell number	N (2)	0	10	Total number of *Giardia* with internal structures
CTOT	Cell number	N (2)	0	10	Total *Cryptosporidium* Counted
CTOTVOL	Cell number/L	N (5.3)	0.000	20.000	Total *Cryptosporidium* per 1 L
CDVQ	n/a	C (1)	n/a	n/a	*Cryptosporidium* Data Value Qualifier (a through z)
CSPOR	Cell number	N (2)	0	10	Total *Cryptosporidium* with Sporozoites counted
CEMPTY	Cell number	N (2)	0	10	Total Empty *Cryptosporidium* Counted
CAMOR	Cell number	N (2)	0	10	Total Amorphous *Cryptosporidium* Counted
CNUC	Cell number	N (2)	0	10	Total *Cryptosporidium* with nuclei counted

Table Name, cont'd: CEC ENVIRONMENTAL SAMPLE

Attribute	Units	Format	Minimum	Maximum	Description
CNUC	Cell number	N (2)	0	10	Total *Cryptosporidium* with nuclei counted
TOTCINTS	Cell number	N (2)	0	10	Total number of *Cryptosporidium* with internal structures
ALANA_DT	n/a	mm/dd/yyyy	n/a	n/a	AL Date Analyzed
ALTPIMV	mL	N (6.3)	1.000	100.000	AL Total pre-IMS Volume
ALSAMVOL	μL	N (6.3)	1.000	100.000	Actual Volume Analyzed for Algae
ALEQVOL	L	N (6.3)	0.001	10.000	Equivalent Volume Analyzed for Algae
ALFIELD	Field	N (3)	1	999	Algae Number Field Counted
ALTFIELD	Field	N (6)	1	$1x10^5$	Total Algae Number Fields
ALAC	Cell number	N (4)	1	1000	Algae Count
ALTAC	Cell number	N (7)	1	$1x10^6$	Total Algae Count
ALTACV	Cell number/L	N (9)	1	$1x10^8$	Total Algae per 1 L
ALDVQ	n/a	C (1)	n/a	n/a	Algae Data Value Qualifier (a through z)

Figure E.8 – CEC Environmental Sample Entity

Table Name: AW SAMPLE CC-PCR

Attribute	Units	Format	Minimum	Maximum	Description
SAMP_ID	n/a	C (11)	n/a	n/a	Sample ID
SAMP_DT	n/a	mm/dd/yyyy	n/a	n/a	Sample Date
CCREC_DT	n/a	mm/dd/yyyy	n/a	n/a	Date Received
CCDIS_DT	n/a	mm/dd/yyyy	n/a	n/a	Date Dissociated
CC1ANAIN	n/a	C (5)	n/a	n/a	Analyst Initials
CC_DT	n/a	mm/dd/yyyy	n/a	n/a	CC-PCR Date
CCRES	n/a	C (10) Positive, negative	n/a	n/a	CC-PCR Result
CC2ANAIN	n/a	C (5)	n/a	n/a	Analyst Initials
CCDVQ	n/a	C (1)	n/a	n/a	CC-PCR Qualifier (a through z)

Figure E.9 – AW Sample CC-PCR Entity

Table Name: AW SAMPLE *CLOSTRIDIUM*

Attribute	Units	Format	Minimum	Maximum	Description
SAMP_ID	n/a	C (11)	n/a	n/a	Sample ID
SAMP_DT	n/a	mm/dd/yyyy	n/a	n/a	Sample Date
CLREC_DT	n/a	mm/dd/yyyy	n/a	n/a	CL Date Received AW
CLFIL_DT	n/a	mm/dd/yyyy	n/a	n/a	CL Date Filtered
CLVOL	mL	N (5.2)	0.500	500.00	CL Volume Filtered
CLCFU	CFU	N (4)	0	1000	*Clostridium perfringens* Count
CLCFUVOL	CFU/100 mL	N (6.2)	0.00	1000.00	*Clostridium perfringens* per 100 mL
CLANAIN	n/a	C (5)	n/a	n/a	CL Analyst Initials
CLDVQ	n/a	C (1)	n/a	n/a	CL Data Value Qualifier (a through z)

Figure E.10 – AW Sample *Clostridium* Entity

Table Name: AW SAMPLE PHAGE

Attribute	Units	Format	Minimum	Maximum	Description
SAMP_ID	n/a	C (11)	n/a	n/a	Sample ID
SAMP_DT	n/a	mm/dd/yyyy	n/a	n/a	Sample Date
PGREC_DT	n/a	mm/dd/yyyy	n/a	n/a	PG Date Received AW
PG_DT	n/a	mm/dd/yyyy	n/a	n/a	PG Date Sample Assayed by SAL method
PGTVOLAY	L	N (5.2)	0.00	100.00	PG Total Volume Assayed
CNCFUVOL	PFU/100 mL	N (6.2)	0.00	1000.00	Mean Somatic Coliphage
FACFUVOL	PFU/100 mL	N (6.2)	0.00	1000.00	Mean Male Specific Coliphage
PGANAIN	n/a	C (5)	n/a	n/a	PGAnalyst Initials
PGDVQ	n/a	C (1)	n/a	n/a	PG Data Value Qualifier (a through z)

Figure E.11 – AW Sample Phage Entity

Table Name: AW SAMPLE *BACILLUS*

Attribute	Units	Format	Minimum	Maximum	Description
SAMP_ID	n/a	C (11)	n/a	n/a	Sample ID
SAMP_DT	n/a	mm/dd/yyyy	n/a	n/a	Sample Date
BSREC_DT	n/a	mm/dd/yyyy	n/a	n/a	BS Date Received AW
BSFIL_DT	n/a	mm/dd/yyyy	n/a	n/a	BS Date Filtered
BSAMOUT	mL	N (6.3)	0.500	500.000	BS Amount Plated Out (mL)
BSVOL	mL	N (6.3)	0.500	500.000	BS Volume Filtered
BSCFU	CFU	N (4)	0	1000	*Bacillus subtilis* Count
BSCFUVOL	CFU/100 mL	N (6.2)	0.00	1000.00	*Bacillus subtilis* per 100 mL
BSANAIN	n/a	C (5)	n/a	n/a	BS Analyst Initials
BSDVQ	n/a	C (1)	n/a	n/a	BS Data Value Qualifier (a through z)

Figure E.12 – AW Sample *Bacillus* Entity

Table Name: AW SAMPLE VIRUS

Attribute	Units	Format	Minimum	Maximum	Description
SAMP_ID	n/a	C (11)	n/a	n/a	Sample ID
SAMP_DT	n/a	mm/dd/yyyy	n/a	n/a	Sample Date
CVREC_DT	n/a	mm/dd/yyyy	n/a	n/a	CV Date Received AW
CVELT_DT	n/a	mm/dd/yyyy	n/a	n/a	CV Date Eluted
CVCON_DT	n/a	mm/dd/yyyy	n/a	n/a	CV Date Concentrated
CVVOL	L	N (5.2)	0.500	500.00	CV Volume of Water Filtered
CVFVOL	mL	N (5.2)	0.500	500.00	CV Final Volume of Sample Concentrate
CVAYVOL	mL	N (5.2)	0.500	500.00	CV Assay Sample Volume
CVOGVOL	L	N (5.2)	0.500	500.00	CV Volume of Original Water Sample Assayed
CVAY_DT	n/a	mm/dd/yyyy	n/a	n/a	CV Date Assayed
CVFLG	n/a	C (1)	n/a	n/a	CV Flag
CVMPN	MPN/100 L	N (6.2)	0	3000.00	CV MPN/100 L
BSANAIN	n/a	C (5)	n/a	n/a	CV Analyst Initials
BSDVQ	n/a	C (1)	n/a	n/a	CV Data Value Qualifier (a through z)

Figure E.13 – AW Sample Virus Entity

APPENDIX F
AVERAGE DAILY FILTER TURBIDITY PLOTS

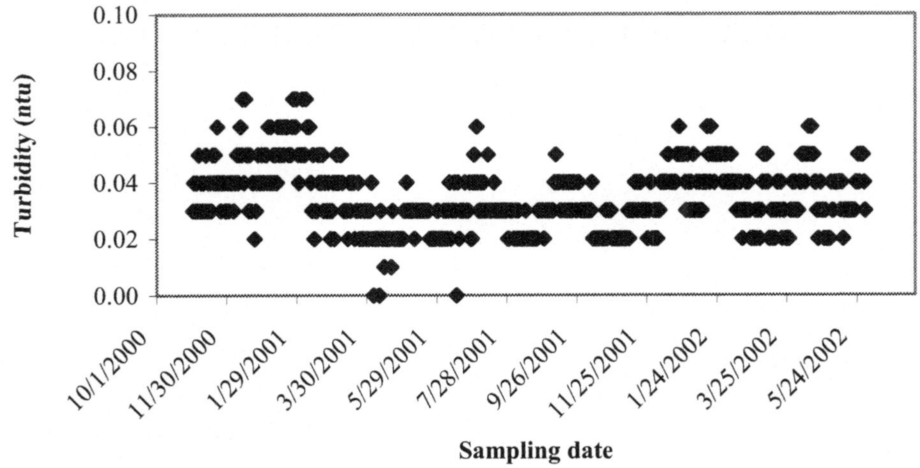

Figure F.1 – Daily turbidity for Filter 1

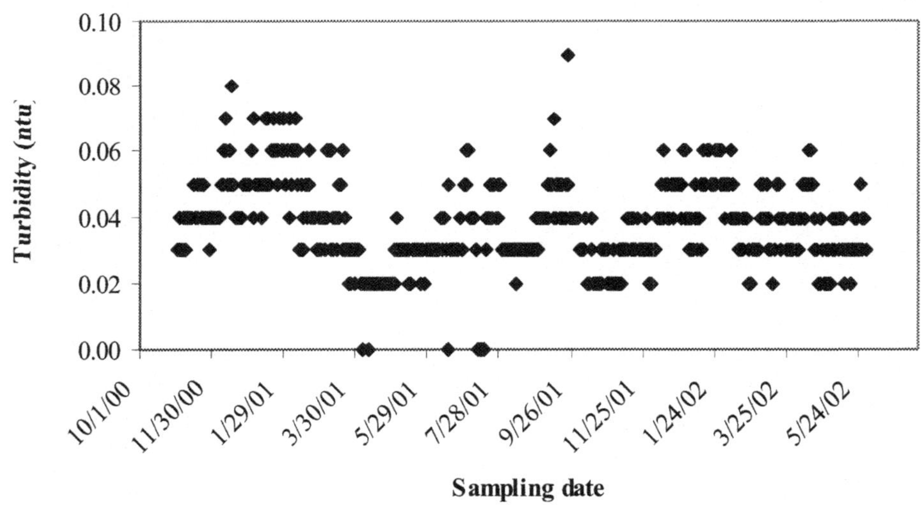

Figure F.2 – Daily turbidity for Filter 2 (the measurement of 0.13 ntu on 1/19/01 is beyond the y-axis scale)

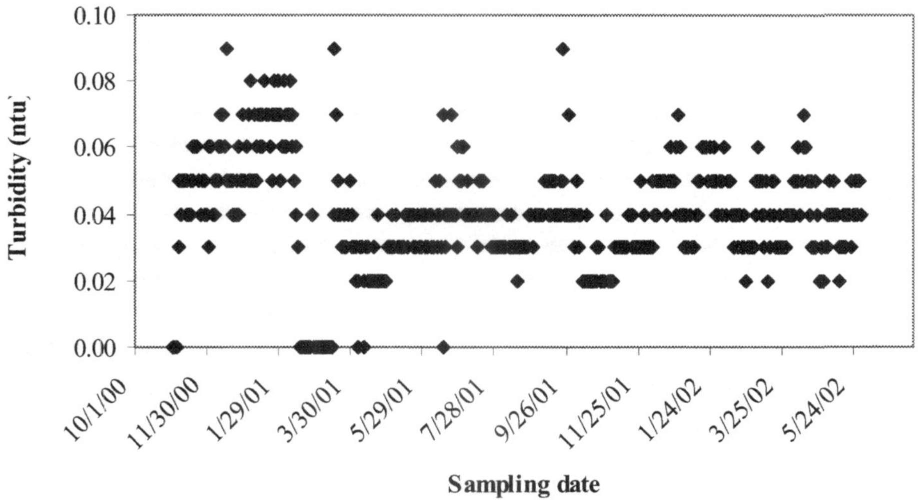

Figure F.3 – Daily turbidity for Filter 3 (the turbidity of 0.16 ntu on 12/20/00 is beyond the y-axis scale)

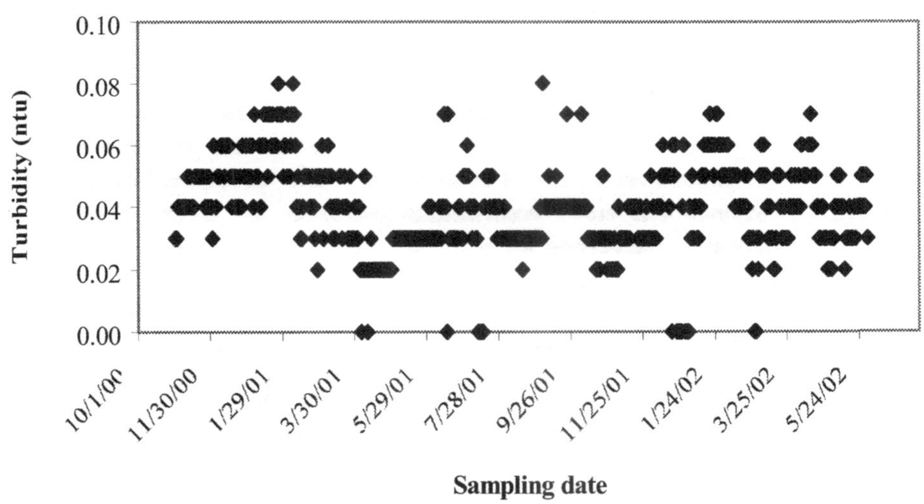

Figure F.4 – Daily turbidity for Filter 4 (the turbidity of 0.42 ntu on 12/4/00 is beyond the y-axis scale)

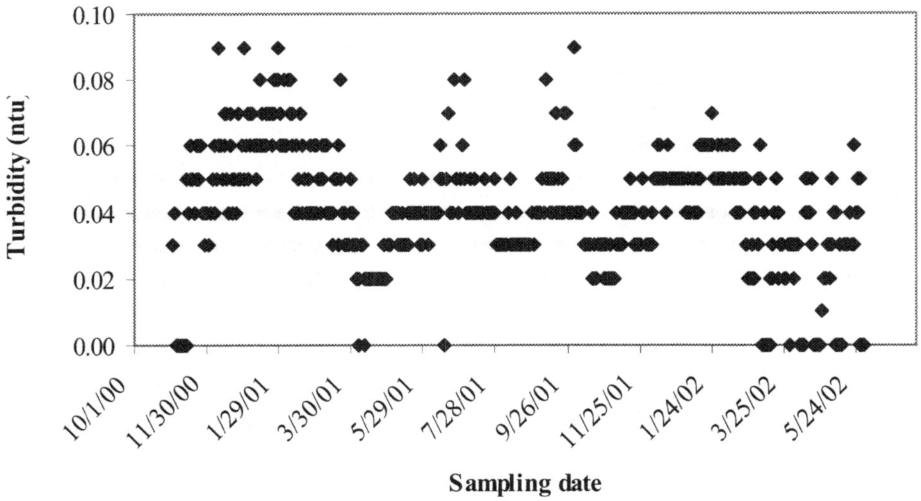

Figure F. 5 – Daily turbidity for Filter 5 (the turbidity of 0.38 ntu on 11/27/00 is beyond the y-axis scale)

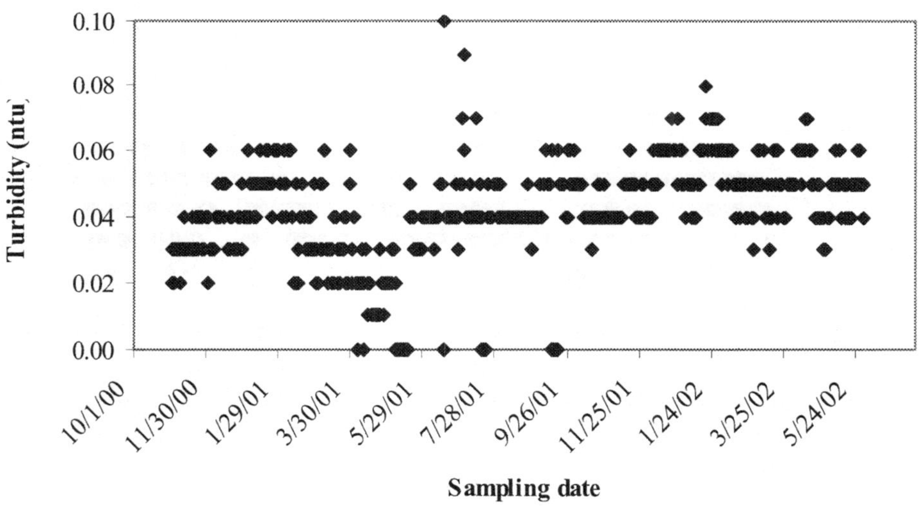

Figure F.6 – Daily turbidity for Filter 6

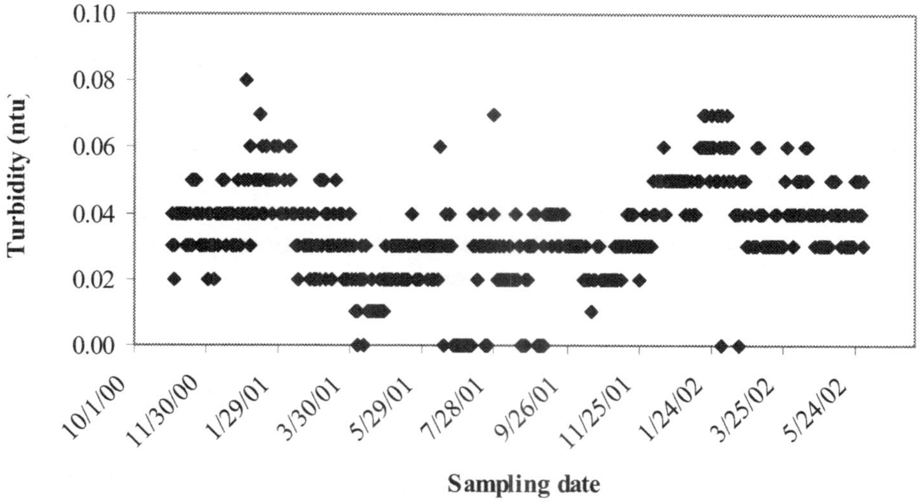

Figure F.7 – Daily turbidity for Filter 7 (the measurement of 0.11 ntu on 6/16/01 is beyond the y-axis scale)

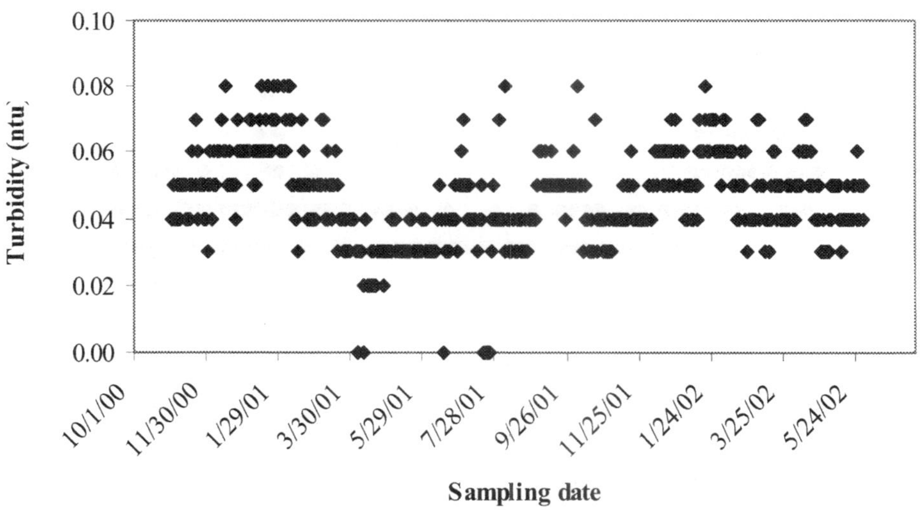

Figure F.8 – Daily turbidity for Filter 8 (the measurements of 0.14 ntu (12/19/00) and 0.12 ntu (6/16/01) are beyond the y-axis scale)

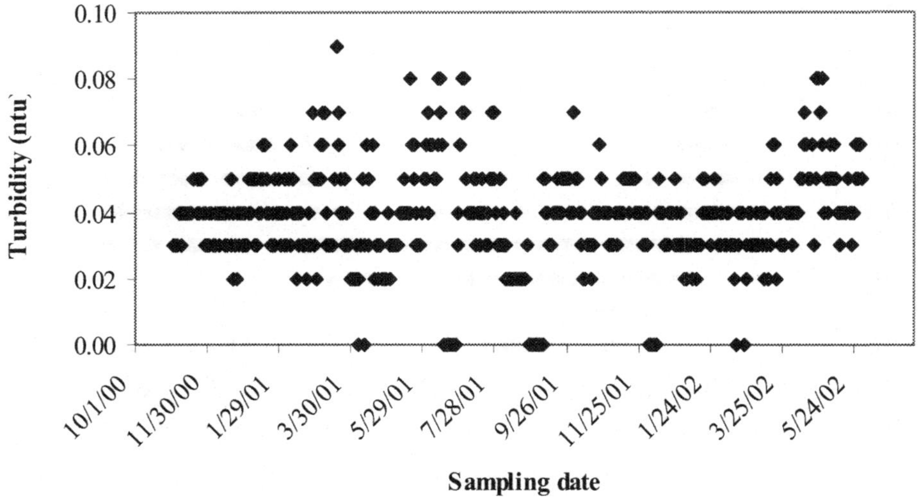

Figure F.9 – Daily turbidity for Filter 9 (the measurement of 0.16 ntu on 11/20/00 is beyond the y-axis scale)

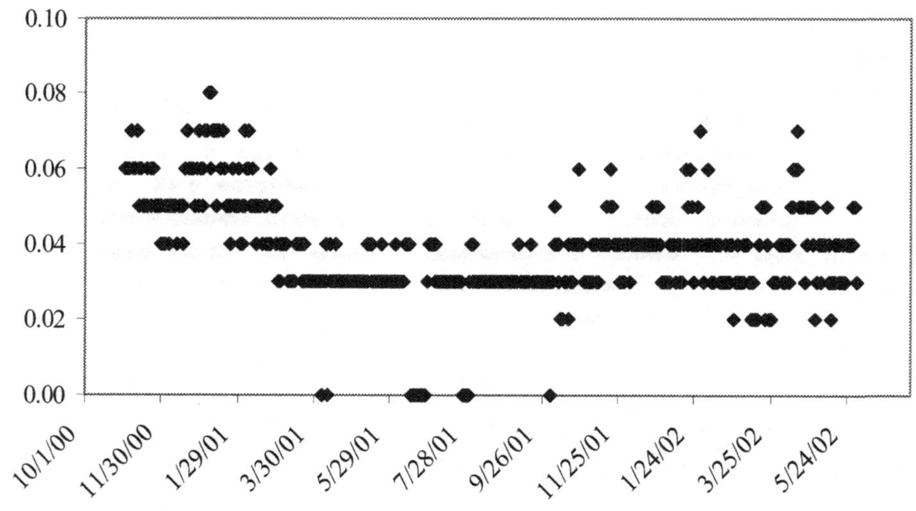

Figure F.10 – Daily turbidity for Filter 10 (the measurement of 0.16 ntu on 11/20/00 is beyond the y-axis scale)

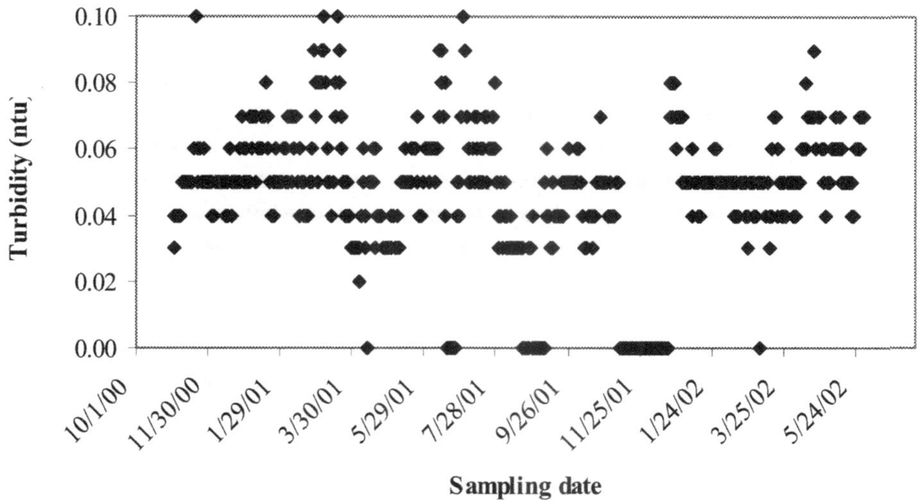

Figure F.11 – Daily turbidity for Filter 11

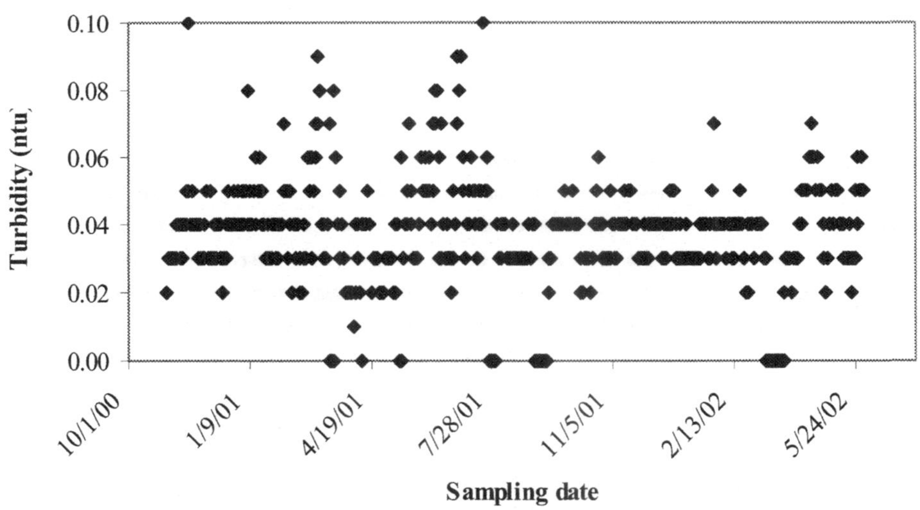

Figure F.12 – Daily turbidity for Filter 12

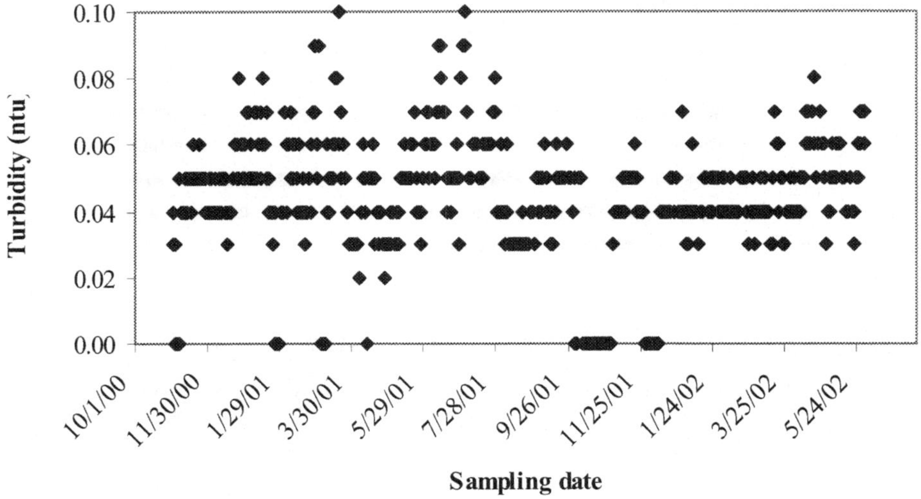

Figure F.13 – Daily turbidity for Filter 13 (the measurements of 0.12 ntu (2/2/01) and 0.14 ntu (7/26/01) are beyond the y-axis scale)

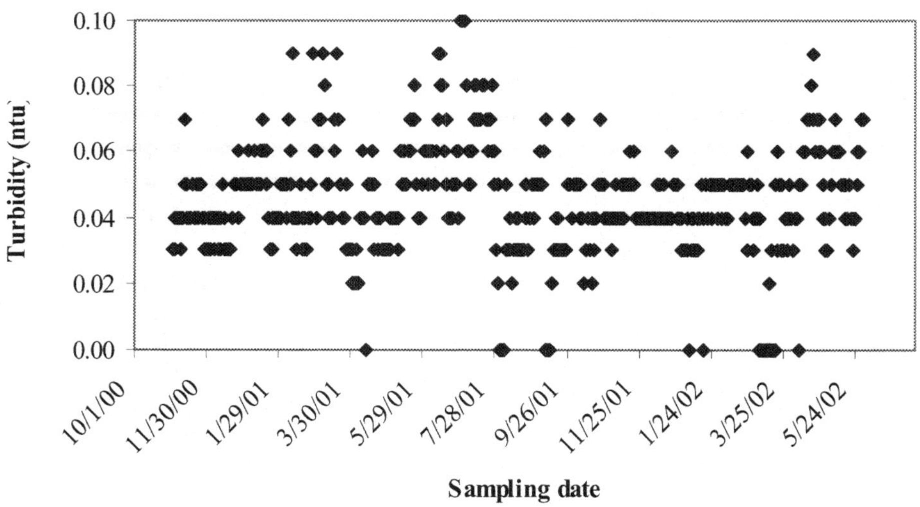

Figure F.14 – Daily turbidity for Filter 14 (the measurements of 0.15 ntu (12/19/00) and 0.11 ntu (7/1/01) are beyond the y-axis scale)

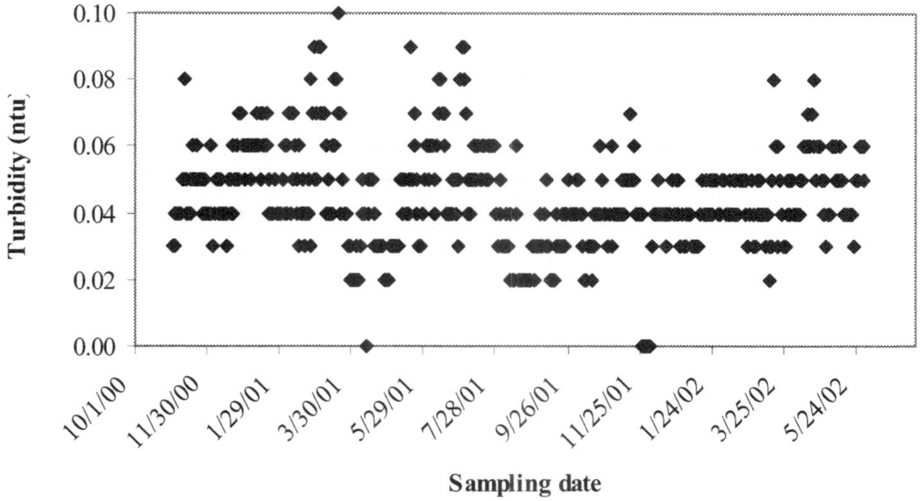

Figure F.15 – Daily turbidity for Filter 15 (the measurement of 0.11 ntu on 3/37/01 is beyond the y-axis scale)

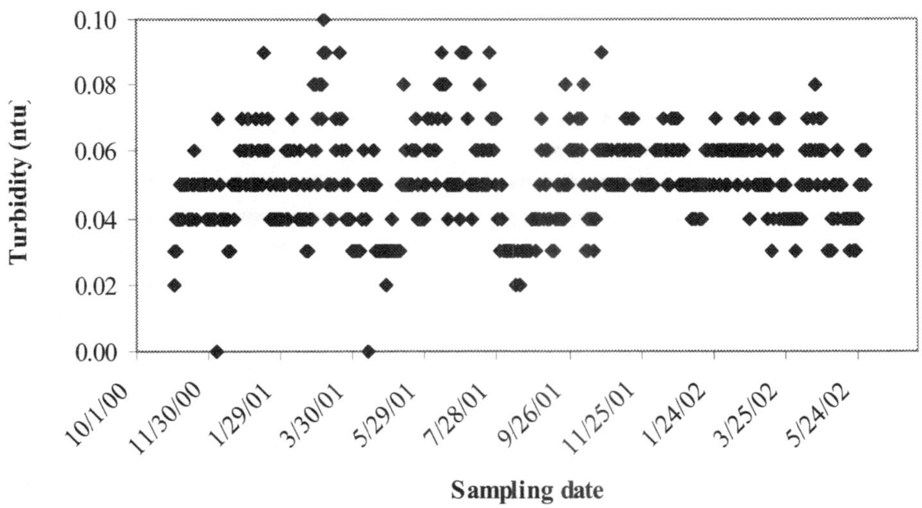

Figure F.16 – Daily turbidity for Filter 16 (the reading of 0.11 ntu measured on 3/18/01 is beyond the y-axis scale)

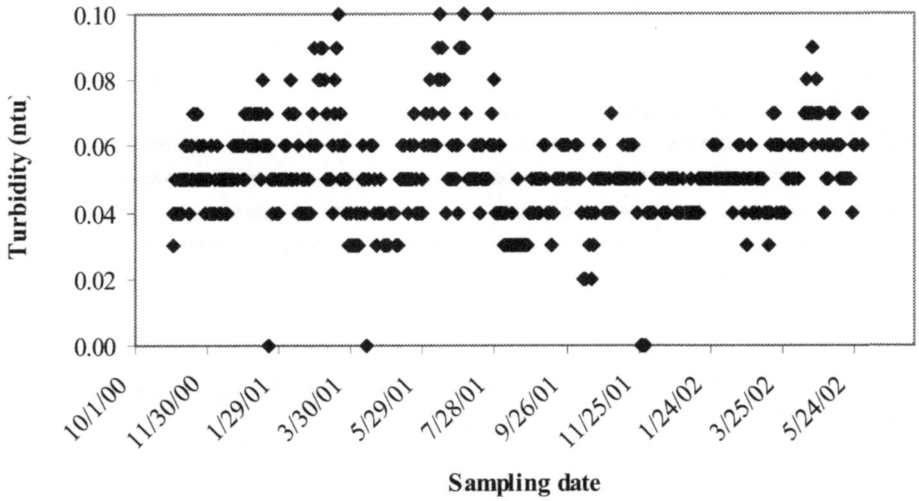

Figure F.17 – Daily turbidity for Filter 17 (the reading of 0.12 ntu measured on 3/7/01 is beyond the y-axis scale)

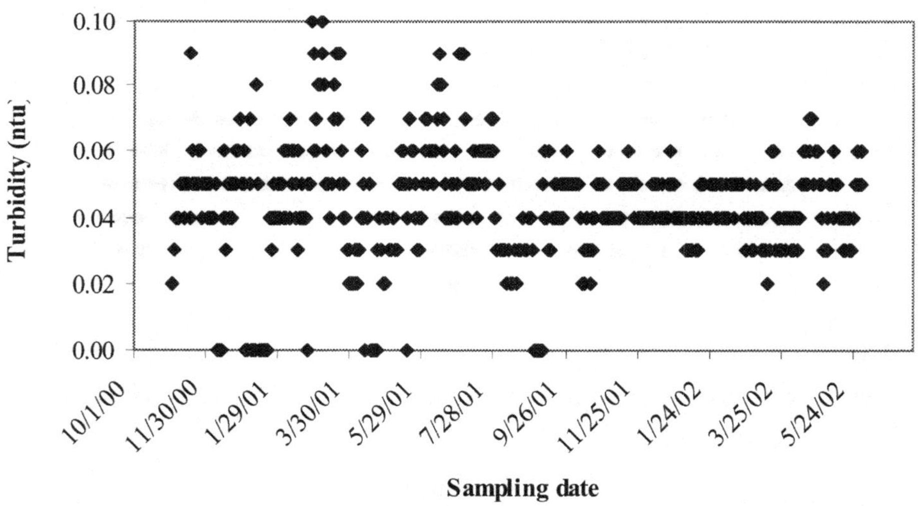

Figure F.18 – Daily turbidity for Filter 18 (the measurement of 0.40 ntu on 11/14/00 is beyond the y-axis scale)

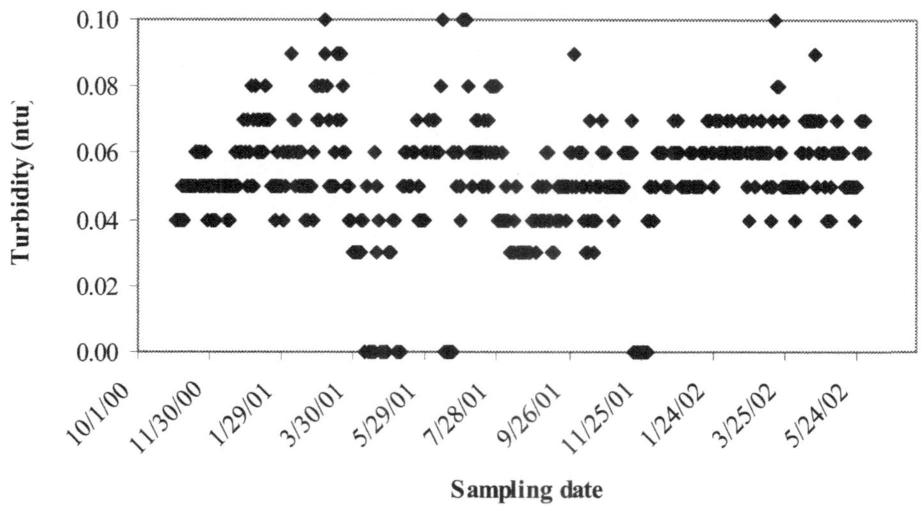

Figure F.19 – Daily turbidity for Filter 19 (the measurement of 0.11 ntu on 6/13/01 and 7/2/01 is beyond the y-axis scale)

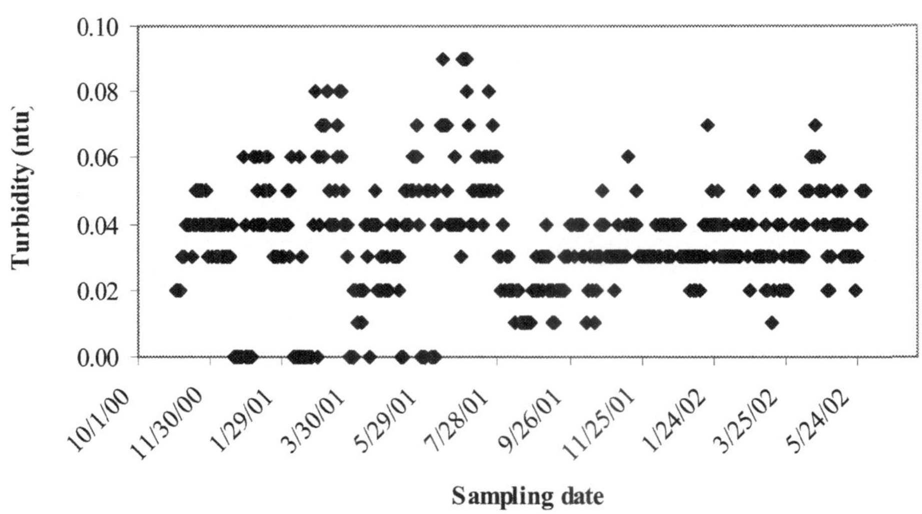

Figure F.20 – Daily turbidity for Filter 20

APPENDIX G
DISTRIBUTION SYSTEM PRESSURE DATA

Location No. = **01**

Starting Date & Time	Ending Date & Time	Elapsed Time (hours)	Minimum psi	Maximum psi	Average psi
03/19/2001 15:00	03/22/2001 13:06	70.1	49.77	77.35	73.70
03/23/2001 9:36	03/30/2001 8:34	167.0	43.94	84.61	73.69
03/30/2001 10:49	04/06/2001 9:49	167.0	43.37	85.89	72.15
04/06/2001 11:46	04/12/2001 8:42	140.9	41.38	94.13	74.63
04/12/2001 15:23	04/19/2001 9:19	161.9	44.22	91.72	- - -
04/19/2001 14:10	04/26/2001 4:36	158.4	52.75	85.46	74.47
04/26/2001 10:33	05/03/2001 7:10	164.6	38.11	88.16	73.83
05/03/2001 9:49	05/09/2001 12:17	146.5	39.67	90.86	73.08
05/09/2001 13:47	05/16/2001 8:44	162.9	43.23	94.42	73.40
05/16/2001 13:21	05/23/2001 6:50	161.5	34.84	92.43	74.18
05/23/2001 13:37	05/28/2001 6:29	112.9	35.69	93.42	73.48
05/30/2001 14:06	06/04/2001 6:58	112.9	44.36	93.99	73.77
06/04/2001 16:16	06/09/2001 9:08	112.9	34.13	92.00	73.06
06/11/2001 14:43	06/16/2001 7:35	112.9	29.58	94.99	73.76
06/18/2001 10:06	06/25/2001 9:04	167.0	33.56	93.99	73.95
06/25/2001 9:24	07/02/2001 10:27	169.0	31.57	89.44	74.13
07/02/2001 10:42	07/09/2001 15:13	172.5	31.14	96.27	74.49
07/09/2001 16:27	07/17/2001 13:53	189.4	32.14	94.99	74.03
07/17/2001 15:32	07/24/2001 13:40	166.1	31.14	94.99	74.17
07/24/2001 15:11	07/31/2001 15:20	168.1	35.69	92.71	74.45
07/31/2001 15:32	08/07/2001 13:33	166.0	30.86	94.42	74.67
08/07/2001 15:21	08/14/2001 9:58	162.6	33.56	94.70	74.89
08/14/2001 15:48	08/22/2001 15:45	192.0	32.85	93.14	74.09
08/22/2001 15:51	08/29/2001 14:13	166.4	37.11	92.43	74.29
08/29/2001 14:33	09/05/2001 10:56	164.4	35.69	93.00	74.60
09/05/2001 11:11	09/13/2001 8:14	189.0	36.83	92.00	73.52
09/13/2001 8:43	09/19/2001 14:45	150.0	37.68	92.00	74.08
09/19/2001 14:48	09/26/2001 13:28	166.7	36.40	90.72	74.40
09/26/2001 15:29	10/02/2001 9:02	137.5	36.69	93.99	74.76
10/02/2001 9:21	10/10/2001 11:31	194.2	35.83	92.71	73.68
10/10/2001 11:40	10/17/2001 13:57	170.3	41.95	86.45	73.54
10/17/2001 15:55	10/25/2001 13:05	189.2	39.10	87.17	73.18
10/25/2001 13:37	11/01/2001 10:40	165.0	43.23	92.71	73.82
11/01/2001 10:55	11/08/2001 11:00	168.1	44.22	92.43	73.44
11/08/2001 11:09	11/14/2001 14:36	147.4	43.22	84.32	73.32
11/14/2001 15:03	11/20/2001 11:00	140.0	43.94	86.45	72.88
11/20/2001 11:07	11/27/2001 8:47	165.7	46.21	82.33	73.06
11/27/2001 10:40	12/04/2001 9:10	166.5	35.12	92.00	73.14
12/04/2001 15:54	12/11/2001 15:26	167.5	43.37	87.45	72.84
12/11/2001 15:41	12/19/2001 10:55	187.2	67.54	79.34	73.31
12/19/2001 15:47	12/27/2001 8:51	185.1	38.96	92.14	75.87
12/27/2001 11:22	01/03/2002 8:39	165.3	42.94	90.15	74.26
01/03/2002 15:43	01/10/2002 4:46	157.0	40.38	93.42	75.41
01/10/2002 9:14	01/16/2002 14:31	149.3	72.52	77.92	74.92
01/16/2002 14:45	01/23/2002 13:05	166.3	39.67	89.87	75.42
01/23/2002 15:10	01/30/2002 13:10	166.0	40.95	90.72	75.38
01/30/2002 14:53	02/06/2002 13:19	166.4	71.52	76.79	74.87
02/06/2002 14:59	02/13/2002 12:21	165.4	71.52	77.07	75.37
02/13/2002 15:14	02/20/2002 12:43	165.5	71.52	79.34	75.29
02/20/2002 14:52	02/27/2002 7:44	160.9	69.82	76.79	75.32
02/27/2002 16:07	03/06/2002 7:43	159.6	71.52	76.79	76.10
03/06/2002 11:28	03/13/2002 12:13	168.8	69.82	78.06	74.86
03/13/2002 13:08	03/20/2002 13:19	168.2	69.82	80.06	75.46
03/20/2002 13:37	03/25/2002 14:33	120.9	42.09	96.27	77.33
03/25/2002 14:43	04/02/2002 14:08	191.4	72.38	82.90	76.60
04/02/2002 15:43	04/09/2002 13:13	165.5	46.64	85.17	76.07
04/09/2002 16:10	04/16/2002 13:58	165.8	49.48	88.16	76.34
04/16/2002 14:02	04/23/2002 13:40	167.6	45.22	91.15	76.18
04/23/2002 14:04	04/30/2002 10:41	164.6	68.54	81.34	75.39
04/30/2002 13:20	05/07/2002 13:30	168.2	71.52	79.63	75.88
05/07/2002 14:30	05/15/2002 12:29	190.0	39.39	87.59	76.02
05/15/2002 15:04	05/22/2002 10:43	163.6	45.36	93.00	75.89
05/22/2002 11:28	05/29/2002 11:19	167.9	42.94	90.44	76.18
05/29/2002 12:16	06/03/2002 10:50	118.6	41.38	90.86	75.29
		0.0			
		SUMMARY:	29.58	96.27	
			MIN	MAX	

Note:
The pressure at Site #1 was oddly steady during the week of 1/10-16/02.
A review of operational and pumping records shoed that the reason for the relatively steady water pressure at this site is that Main Service Pump #12 was run continuously from 5:00 AM, January 10 through January 22 without pump changes being made on Main Service. Main Service Pump #12 pumps at a rate of 4.5 MGD. Prior to 5:00 AM, January 10th different Main Service pumps were sometimes operated and #12 would be turned off. The Operator reports that starting on January 22, #12 pump has been turned off and on, and other larger Main Service pumps sometimes used.

Location No. = **02**

Starting Date & Time	Ending Date & Time	Elapsed Time (hours)	Minimum psi	Maximum psi	Average psi
03/19/2001 14:10	03/22/2001 14:30	72.3	38.96	65.55	53.73
03/22/2001 14:30	03/29/2001 15:03	168.6	39.25	61.43	52.88
03/29/2001 15:03	04/05/2001 15:20	168.3	40.24	62.42	53.57
04/05/2001 15:20	04/12/2001 9:35	162.2	37.11	65.55	53.58
04/12/2001 9:38	04/19/2001 6:53	165.3	43.23	58.44	52.26
04/19/2001 13:44	04/25/2001 10:12	140.5	34.84	63.28	52.79
04/25/2001 10:29	05/02/2001 9:58	167.5	30.00	63.56	51.55
05/02/2001 10:28	05/09/2001 9:07	166.6	31.28	62.28	51.51
05/09/2001 10:21	05/16/2001 7:43	165.4	37.97	60.01	52.20
05/16/2001 10:23	05/23/2001 7:40	165.3	36.12	59.44	51.66
05/23/2001 10:21	05/30/2001 7:14	164.9	37.11	59.44	52.69
05/30/2001 10:17	06/06/2001 7:51	165.6	41.24	63.56	52.90
06/06/2001 10:31	06/13/2001 6:13	163.7	38.11	63.56	50.82
06/13/2001 10:23	06/20/2001 10:14	167.9	33.84	65.55	51.49
06/20/2001 11:29	06/27/2001 12:26	168.9	32.14	64.56	50.59
06/27/2001 14:39	07/03/2001 10:15	139.6	33.13	66.97	49.75
07/03/2001 10:34	07/11/2001 12:58	194.4	33.84	65.27	51.09
07/11/2001 15:05	07/18/2001 14:52	167.8	30.00	63.56	48.02
07/18/2001 14:57	07/25/2001 10:10	163.2	30.00	64.56	51.77
07/25/2001 10:22	08/01/2001 15:58	173.6	31.14	67.54	48.93
08/01/2001 16:01	08/08/2001 15:56	167.9	32.14	62.00	49.49
08/08/2001 16:17	08/15/2001 13:32	165.2	33.13	64.56	49.22
08/15/2001 15:00	08/22/2001 16:08	169.1	36.12	62.99	51.33
08/22/2001 16:21	08/29/2001 15:08	166.8	36.12	65.55	51.77
08/29/2001 15:23	09/05/2001 10:25	163.0	31.14	67.26	50.23
09/05/2001 10:34	09/12/2001 14:46	172.2	35.69	64.56	52.52
09/12/2001 16:03	09/19/2001 14:18	166.3	36.12	64.56	52.07
09/19/2001 15:34	09/26/2001 15:32	168.0	42.23	64.56	52.58
09/26/2001 16:13	10/01/2001 16:31	120.3	37.68	62.42	51.77
10/01/2001 16:36	10/09/2001 11:22	186.8	36.12	64.56	52.87
10/09/2001 11:26	10/17/2001 16:24	197.0	34.70	65.98	53.79
10/17/2001 16:39	10/25/2001 12:50	188.2	41.66	64.56	53.40
10/25/2001 13:03	11/01/2001 10:14	165.2	31.57	63.99	53.68
11/01/2001 10:22	11/08/2001 9:29	167.1	29.01	64.56	53.15
11/08/2001 10:17	11/15/2001 9:38	167.3	35.12	64.26	53.02
11/15/2001 10:04	11/21/2001 15:20	149.3	42.23	62.42	52.54
11/21/2001 15:35	11/28/2001 15:22	167.8	40.24	64.56	53.07
11/28/2001 16:02	12/06/2001 10:09	186.1	37.11	64.32	53.23
12/06/2001 10:14	12/13/2001 10:26	168.2	27.59	63.99	52.56
12/13/2001 10:30	12/20/2001 9:55	167.4	-0.85	62.42	52.08
12/20/2001 10:35	12/27/2001 9:13	166.6	34.13	64.56	52.47
12/27/2001 10:36	01/03/2002 10:04	167.5	34.70	60.43	50.54
01/03/2002 10:19	01/10/2002 10:05	167.8	37.68	63.28	51.60
01/10/2002 10:11	01/17/2002 10:14	168.0	39.39	63.56	52.40
01/17/2002 10:28	01/24/2002 11:01	168.5	34.13	65.55	53.62
01/24/2002 11:06	01/31/2002 13:45	170.6	42.23	63.56	53.70
01/31/2002 13:57	02/07/2002 10:51	164.9	32.13	64.56	53.66
02/07/2002 10:55	02/14/2002 10:19	167.4	38.96	64.46	54.74
02/14/2002 10:36	02/21/2002 10:11	167.6	33.13	65.98	54.17
02/21/2002 10:33	02/28/2002 9:33	167.0	43.23	64.56	54.07
02/28/2002 10:16	03/07/2002 10:44	168.5	42.94	64.56	53.15
03/07/2002 11:02	03/14/2002 9:38	166.6	29.86	64.46	54.22
03/14/2002 9:41	03/21/2002 10:43	169.0	33.13	67.26	54.97
03/21/2002 10:58	03/25/2002 13:31	98.6	27.59	76.64	52.97
03/25/2002 13:49	04/02/2002 14:07	192.3	34.13	66.97	54.65
04/02/2002 14:30	04/10/2002 12:34	190.1	35.12	67.54	55.47
04/10/2002 13:38	04/17/2002 11:15	165.6	27.02	66.55	53.56
04/17/2002 11:19	04/24/2002 11:17	168.0	38.11	65.55	54.47
04/24/2002 11:21	05/01/2002 10:34	167.2	42.23	65.55	54.63
05/01/2002 10:44	05/08/2002 10:26	167.7	32.14	66.55	54.05
05/08/2002 10:34	05/15/2002 14:28	171.9	36.12	63.56	53.59
05/15/2002 14:30	05/22/2002 10:45	164.3	36.10	66.55	54.18
05/22/2002 10:55	05/29/2002 13:31	170.6	39.25	65.55	53.78
05/29/2002 13:45	06/03/2002 10:42	117.0	31.14	64.56	52.19
		0.0			
		SUMMARY:	27.02	76.64	
			MIN*	MAX	

* Exceptions noted below

* At Site #2 on Monday 12/17/01, there was a negative pressure event of -0.85 psi from 0:37am until 12:12pm. A maintenance person was changing the mechanical gauge that was there to an electronic gauge for the SCADA system.

Location No. = **03**

Starting Date & Time	Ending Date & Time	Elapsed Time (hours)	Minimum psi	Maximum psi	Average psi
03/15/2001 10:00	03/19/2001 10:35	96.6	33.56	59.44	48.76
03/22/2001 13:16	03/29/2001 11:14	166.0	34.84	59.15	47.83
03/29/2001 11:14	04/05/2001 15:40	172.4	35.41	58.87	47.69
04/05/2001 15:40	04/11/2001 15:25	143.7	34.13	60.15	47.42
04/11/2001 15:28	04/19/2001 6:02	182.6	34.27	59.15	47.60
04/19/2001 10:10	04/25/2001 11:48	145.6	36.40	60.58	48.58
04/25/2001 11:52	05/02/2001 12:56	169.1	33.84	57.87	47.58
05/02/2001 14:29	05/09/2001 11:10	164.7	34.13	58.87	48.52
05/09/2001 11:42	05/15/2001 14:13	146.5	34.27	59.15	48.11
05/15/2001 16:41	05/22/2001 11:43	163.0	17.35	58.87	48.44
05/22/2001 16:27	05/29/2001 11:53	163.4	38.39	60.15	48.54
05/29/2001 18:09	06/05/2001 11:08	161.0	29.72	60.43	48.57
06/05/2001 16:33	06/12/2001 12:49	164.3	29.58	58.87	49.73
06/12/2001 16:10	06/19/2001 13:58	165.8	28.01	60.15	49.31
06/19/2001 14:56	06/26/2001 15:26	168.5	36.54	62.42	49.64
06/26/2001 15:36	07/02/2001 13:56	142.3	36.40	64.13	51.22
07/02/2001 15:06	07/10/2001 16:04	193.0	31.85	62.14	51.59
07/10/2001 16:19	07/17/2001 15:29	167.2	24.17	66.97	53.06
07/17/2001 15:55	07/24/2001 15:11	167.3	31.57	63.13	51.81
07/24/2001 15:25	07/31/2001 14:02	166.6	25.74	62.14	51.58
07/31/2001 15:53	08/07/2001 12:49	164.9	9.53	63.13	51.97
08/07/2001 15:45	08/14/2001 12:42	165.0	27.02	61.71	52.54
08/14/2001 16:04	08/21/2001 10:05	162.0	37.11	62.85	50.89
08/21/2001 16:47	08/28/2001 16:17	167.5	36.40	63.42	50.52
08/28/2001 16:41	09/04/2001 15:28	166.8	18.34	60.15	51.27
09/04/2001 16:39	09/11/2001 14:39	166.0	34.55	62.85	50.59
09/11/2001 17:50	09/18/2001 14:57	165.1	33.56	58.16	50.01
09/18/2001 16:26	09/25/2001 16:58	168.5	31.85	59.86	49.97
09/25/2001 17:23	10/01/2001 14:51	141.5	32.85	60.58	51.31
10/01/2001 16:52	10/09/2001 16:50	192.0	30.00	59.86	50.09
10/09/2001 17:29	10/16/2001 16:29	167.0	37.40	57.16	49.41
10/16/2001 17:07	10/24/2001 7:36	182.5	33.27	59.15	49.93
10/24/2001 11:04	10/31/2001 16:01	173.0	36.12	58.30	49.64
10/31/2001 16:57	11/07/2001 13:57	165.0	30.57	60.43	50.03
11/07/2001 16:03	11/14/2001 14:38	166.6	25.36	58.30	49.56
11/14/2001 14:47	11/20/2001 8:37	137.8	36.83	56.59	48.92
11/20/2001 10:42	11/27/2001 10:14	167.5	36.12	59.15	49.09
11/27/2001 10:23	12/04/2001 15:01	172.6	33.27	57.87	48.69
12/04/2001 16:19	12/11/2001 13:40	165.4	37.40	58.87	49.18
12/11/2001 15:59	12/19/2001 12:56	189.0	37.82	56.88	48.95
12/19/2001 15:30	12/26/2001 16:32	169.0	35.55	57.16	47.84
12/26/2001 19:23	01/02/2002 15:43	164.3	34.84	57.30	47.77
01/02/2002 15:47	01/09/2002 10:49	163.0	35.83	58.20	47.76
01/09/2002 14:59	01/15/2002 12:41	141.7	35.12	57.59	48.04
01/15/2002 16:39	01/22/2002 14:10	165.5	30.57	56.59	47.29
01/22/2002 16:23	01/29/2002 16:33	168.2	34.55	57.30	48.36
01/29/2002 16:42	02/05/2002 11:25	162.7	32.56	58.89	47.58
02/05/2002 15:04	02/12/2002 14:57	167.9	37.68	57.30	47.82
02/12/2002 16:20	02/19/2002 11:29	163.2	31.28	58.30	48.28
02/19/2002 15:34	02/26/2002 15:17	167.7	31.57	58.87	48.19
02/26/2002 16:52	03/05/2002 15:52	167.0	36.40	57.87	47.70
03/05/2002 16:58	03/12/2002 16:29	167.5	35.83	57.16	48.22
03/12/2002 16:32	03/19/2002 17:09	168.6	36.83	59.44	48.54
03/19/2002 17:26	03/25/2002 12:27	139.0	29.58	58.58	48.25
03/25/2002 14:12	04/02/2002 13:05	190.9	32.28	60.15	48.27
04/02/2002 13:14	04/10/2002 11:58	190.7	33.84	58.30	48.69
04/10/2002 13:02	04/17/2002 9:22	164.3	31.85	59.44	49.27
04/17/2002 11:04	04/24/2002 11:33	168.5	35.55	58.20	49.00
04/24/2002 11:36	05/01/2002 10:09	166.6	35.12	61.20	48.19
05/01/2002 10:12	05/08/2002 9:58	167.8	30.29	60.86	48.68
05/08/2002 10:20	05/15/2002 12:47	170.4	35.12	59.86	48.79
05/15/2002 13:55	05/22/2002 10:23	164.5	31.85	63.13	48.76
05/22/2002 10:40	05/29/2002 11:00	168.3	25.74	58.87	49.27
05/29/2002 11:05	06/03/2002 9:02	117.9	37.82	60.86	49.58
		0.0			
		SUMMARY:	9.53	66.97	
			MIN	MAX	

Location No. = **04**

Starting Date & Time	Ending Date & Time	Elapsed Time (hours)	Minimum psi	Maximum psi	Average psi
03/23/2001 11:25	03/30/2001 8:00	164.6	54.03	92.85	66.40
03/30/2001 13:35	04/06/2001 14:15	168.7	43.80	95.70	66.46
04/06/2001 14:15	04/12/2001 14:57	144.7	37.97	92.85	66.62
04/12/2001 15:00	04/20/2001 13:30	190.5	44.08	89.58	65.89
04/20/2001 14:16	04/26/2001 10:47	140.5	50.19	75.36	66.41
04/26/2001 10:55	05/03/2001 10:05	167.2	47.21	94.13	67.68
05/03/2001 10:12	05/09/2001 12:14	146.0	35.41	96.69	66.46
05/09/2001 13:28	05/16/2001 10:03	164.6	56.31	94.84	66.56
05/16/2001 10:46	05/23/2001 10:46	168.0	43.80	96.83	68.49
05/23/2001 11:14	05/30/2001 13:38	170.4	50.19	94.84	67.77
05/30/2001 13:42	06/06/2001 13:18	167.6	39.53	93.85	68.12
06/06/2001 13:36	06/13/2001 11:25	165.8	47.21	94.84	69.28
06/13/2001 12:15	06/20/2001 9:38	165.4	46.92	97.12	68.98
06/20/2001 11:07	06/27/2001 14:10	171.1	48.20	96.69	68.79
06/27/2001 14:16	07/03/2001 11:06	140.8	46.64	95.70	68.42
07/03/2001 13:45	07/11/2001 14:25	192.7	48.20	96.83	66.41
07/11/2001 14:30	07/18/2001 14:20	167.8	34.41	94.84	69.26
07/18/2001 14:34	07/25/2001 12:56	166.4	48.20	96.83	68.41
07/25/2001 13:07	08/01/2001 15:35	170.5	48.20	98.40	68.93
08/01/2001 15:38	08/08/2001 15:52	168.2	43.09	106.79	71.00
08/08/2001 15:57	08/15/2001 14:08	166.2	19.77	98.40	70.83
08/15/2001 14:12	08/22/2001 12:50	166.6	52.19	97.97	70.05
08/22/2001 13:01	08/29/2001 14:38	169.6	55.31	98.97	71.36
08/29/2001 14:44	09/05/2001 10:43	164.0	52.19	97.97	69.98
09/05/2001 10:50	09/12/2001 13:47	171.0	45.64	98.97	69.83
09/12/2001 15:36	09/19/2001 14:51	167.3	52.19	98.97	68.63
09/19/2001 15:02	09/26/2001 14:22	167.3	43.09	96.83	68.21
09/26/2001 15:48	10/01/2001 16:05	120.3	47.21	98.68	68.38
10/01/2001 16:12	10/09/2001 10:47	186.6	51.19	98.97	68.24
10/09/2001 11:00	10/17/2001 15:58	197.0	49.20	98.97	67.84
10/17/2001 16:10	10/25/2001 16:09	192.0	54.03	99.62	68.01
10/26/2001 12:29	11/01/2001 10:34	142.1	55.31	96.83	68.23
11/01/2001 10:39	11/08/2001 10:15	167.6	45.64	95.84	67.53
11/08/2001 10:47	11/15/2001 10:06	167.3	44.08	95.24	67.39
11/15/2001 10:20	11/21/2001 14:30	148.2	54.32	94.84	67.27
11/21/2001 15:51	11/28/2001 15:38	167.8	40.81	78.06	67.92
11/28/2001 15:45	12/06/2001 10:44	187.0	39.53	97.40	67.76
12/06/2001 10:48	12/13/2001 10:11	167.4	55.74	96.69	67.47
12/13/2001 10:52	12/20/2001 10:52	168.0	58.30	96.83	68.00
12/20/2001 10:58	12/27/2001 8:03	165.1	53.18	98.97	70.63
12/27/2001 11:00	01/03/2002 10:31	167.5	58.30	91.86	69.37
01/03/2002 10:50	01/10/2002 10:37	167.8	56.31	78.63	68.93
01/10/2002 10:42	01/17/2002 10:27	167.8	57.87	75.65	69.02
01/17/2002 10:57	01/24/2002 9:49	166.9	54.32	76.64	68.99
01/24/2002 11:20	01/31/2002 13:41	170.4	52.75	76.68	69.17
01/31/2002 14:11	02/07/2002 10:45	164.6	56.31	75.08	68.52
02/07/2002 11:10	02/14/2002 10:43	167.5	42.80	94.84	69.14
02/14/2002 10:50	02/21/2002 10:47	168.0	58.30	100.96	70.06
02/21/2002 11:00	02/28/2002 9:21	166.3	55.31	97.97	69.48
02/28/2002 10:34	03/07/2002 11:01	168.4	51.19	95.41	69.54
03/07/2002 11:18	03/14/2002 9:52	166.6	57.02	74.08	68.92
03/14/2002 10:03	03/21/2002 10:34	168.5	58.30	99.39	69.48
03/21/2002 11:14	03/25/2002 14:20	99.1	58.30	97.40	70.85
03/25/2002 14:29	04/02/2002 14:41	192.2	57.87	92.85	70.10
04/02/2002 14:45	04/10/2002 12:15	189.5	54.75	90.86	69.51
04/10/2002 13:20	04/17/2002 11:29	166.2	51.76	99.96	69.83
04/17/2002 11:34	04/24/2002 10:53	167.3	52.19	94.84	69.57
04/24/2002 11:05	05/01/2002 11:04	168.0	58.30	95.41	69.25
05/01/2002 11:10	05/08/2002 10:15	167.1	58.30	94.84	69.33
05/08/2002 11:00	05/15/2002 14:18	171.3	52.19	94.42	69.54
05/15/2002 14:48	05/22/2002 10:51	164.0	58.30	98.40	70.05
05/22/2002 11:11	05/29/2002 11:34	168.4	56.74	94.84	70.28
05/29/2002 11:37	06/03/2002 10:55	119.3	52.19	97.97	71.52
		0.0			
		SUMMARY:	19.77	106.79	
			MIN	MAX	

Location No. = **05**

Starting Date & Time	Ending Date & Time	Elapsed Time (hours)	Minimum psi	Maximum psi	Average psi
03/16/2001 12:36	03/21/2001 22:32	129.9	44.36	91.72	50.01
03/23/2001 10:10	03/30/2001 7:35	165.4	46.78	92.00	50.81
03/30/2001 12:04	04/06/2001 13:10	169.1	46.36	91.00	51.12
04/06/2001 13:10	04/12/2001 8:35	139.4	47.64	85.46	49.82
04/12/2001 8:38	04/20/2001 6:44	190.1	49.63	84.75	52.03
04/20/2001 10:13	04/26/2001 9:35	143.4	49.63	92.28	53.50
04/26/2001 10:03	05/03/2001 6:55	164.9	50.05	83.18	51.60
05/03/2001 9:10	05/10/2001 6:55	165.8	55.17	92.71	57.04
05/10/2001 11:37	05/16/2001 3:54	136.3	78.78	92.71	83.92
05/16/2001 12:21	05/23/2001 6:17	161.9	1.42	98.54	85.96
05/23/2001 12:37	05/30/2001 10:55	166.3	50.05	93.28	86.24
05/30/2001 11:58	06/06/2001 6:19	162.3	51.05	86.03	81.93
06/06/2001 11:52	06/13/2001 8:06	164.2	78.92	86.03	85.72
06/13/2001 12:50	06/20/2001 6:06	161.3	86.03	86.03	86.03
06/20/2001 12:01	06/27/2001 10:16	166.2	86.03	98.54	88.56
06/27/2001 15:01	07/03/2001 5:25	134.4	90.29	90.29	90.29
07/03/2001 12:31	07/11/2001 11:51	191.3	90.58	102.52	98.31
07/11/2001 15:35	07/18/2001 8:51	161.3	102.52	102.52	102.52
07/18/2001 15:18	07/25/2001 8:35	161.3	102.52	102.52	102.52
07/25/2001 11:07	08/02/2001 7:17	188.2	102.52	102.52	102.52
08/02/2001 9:02	08/09/2001 2:19	161.3	102.52	102.52	102.52
08/09/2001 8:45	08/16/2001 2:02	161.3	102.52	102.52	102.52
08/16/2001 10:35	08/23/2001 6:55	164.3	73.94	95.27	78.43
08/23/2001 11:11	08/30/2001 9:24	166.2	84.18	88.30	86.51
08/30/2001 10:09	09/06/2001 2:47	160.6	84.18	92.28	87.73
09/06/2001 9:42	09/13/2001 1:49	160.1	88.30	96.27	91.32
09/13/2001 9:12	09/20/2001 7:51	166.6	64.13	99.25	77.59
09/20/2001 9:48	09/27/2001 8:16	166.5	99.25	103.38	101.74
09/27/2001 10:27	10/02/2001 6:56	116.5	99.25	99.25	99.25
10/02/2001 9:43	10/10/2001 5:52	188.1	99.25	99.25	99.25
10/10/2001 10:40	10/18/2001 5:07	186.4	74.51	101.24	96.73
10/18/2001 9:56	10/25/2001 6:59	165.0	57.45	90.72	78.21
10/25/2001 12:23	11/01/2001 2:38	158.3	62.38	94.84	79.18
11/01/2001 9:57	11/08/2001 5:33	163.6	64.32	96.65	76.88
11/08/2001 9:48	11/15/2001 5:06	163.3	68.23	99.11	80.71
11/15/2001 9:41	11/21/2001 9:02	143.3	67.48	99.23	79.77
11/21/2001 15:06	11/28/2001 8:45	161.7	68.92	91.00	82.29
11/28/2001 12:42	12/05/2001 7:09	162.5	65.98	93.00	79.95
12/05/2001 13:22	12/12/2001 9:33	164.2	56.45	84.46	69.06
12/12/2001 15:24	12/19/2001 10:29	163.1	51.33	88.45	65.31
12/19/2001 13:30	12/26/2001 10:59	165.5	51.05	91.29	63.07
12/26/2001 18:56	01/02/2002 14:01	163.1	48.35	76.59	62.49
01/02/2002 16:08	01/09/2002 13:04	164.9	47.35	90.01	53.88
01/09/2002 14:32	01/16/2002 14:00	167.5	46.78	87.45	52.13
01/16/2002 14:20	01/23/2002 13:45	167.4	48.35	90.01	51.73
01/23/2002 14:46	01/30/2002 7:09	160.4	46.78	91.00	52.18
01/30/2002 14:29	02/06/2002 15:14	168.8	44.36	92.57	50.91
02/06/2002 15:21	02/13/2002 14:05	166.7	45.36	91.72	51.63
02/13/2002 14:33	02/20/2002 10:49	164.3	44.65	91.00	51.16
02/20/2002 14:30	02/27/2002 6:30	160.0	46.07	90.01	51.69
02/27/2002 12:46	03/06/2002 5:10	160.4	47.64	83.75	50.88
03/06/2002 11:47	03/13/2002 5:48	162.0	46.78	91.72	51.11
03/13/2002 13:35	03/20/2002 10:29	164.9	47.07	92.71	51.37
03/20/2002 14:03	03/26/2002 8:41	138.6	46.07	90.29	52.09
03/26/2002 8:57	04/02/2002 12:31	171.6	48.35	92.57	51.86
04/02/2002 13:37	04/09/2002 12:56	167.3	47.35	91.72	52.38
04/09/2002 14:29	04/16/2002 12:48	166.3	44.79	87.73	51.94
04/16/2002 13:38	04/23/2002 11:20	165.7	49.63	90.58	53.40
04/23/2002 13:38	04/30/2002 9:43	164.1	52.90	85.03	56.52
04/30/2002 13:40	05/07/2002 9:56	164.3	52.90	92.71	73.16
05/07/2002 13:53	05/15/2002 12:15	190.4	55.88	94.56	73.03
05/15/2002 13:37	05/22/2002 4:57	159.3	74.51	95.84	87.61
05/22/2002 13:45	05/29/2002 9:56	164.2	77.73	95.84	90.82
05/29/2002 12:37	06/03/2002 7:31	114.9	74.84	99.82	88.94
		0.0			
		SUMMARY:	44.36 MIN*	103.38 MAX	

* Exceptions noted below

* The May 21, 2001 low pressure (1.42 psi) incident at site #5 (a hotel) occurred because one of the guests put a shirt on one of the sprinkler heads in the room and pulled the shirt down and the sprinkler head with it. The water to the hotel was turned off.

Location No. = **06**

Starting Date & Time	Ending Date & Time	Elapsed Time (hours)	Minimum psi	Maximum psi	Average psi
03/15/2001 15:30	03/18/2001 7:35	64.1	68.40	82.05	75.58
03/22/2001 14:03	03/29/2001 14:33	168.5	69.25	77.35	75.64
03/29/2001 14:33	04/05/2001 14:30	168.0	39.53	79.91	75.17
04/05/2001 14:30	04/11/2001 14:52	144.4	69.96	80.62	75.16
04/11/2001 14:55	04/19/2001 6:06	183.2	68.40	83.18	75.03
04/19/2001 9:19	04/25/2001 6:11	140.9	66.69	81.90	75.38
04/25/2001 13:06	05/02/2001 10:41	165.6	39.25	82.62	74.55
05/02/2001 12:53	05/09/2001 8:52	164.0	57.59	82.62	74.96
05/09/2001 12:36	05/15/2001 15:13	146.6	48.06	81.62	74.72
05/15/2001 15:57	05/22/2001 10:42	162.8	37.40	86.31	73.87
05/22/2001 16:05	05/30/2001 11:05	187.0	45.50	82.05	75.24
05/30/2001 11:37	06/06/2001 7:30	163.9	40.53	83.47	75.74
06/06/2001 12:11	06/13/2001 12:38	168.4	55.60	86.74	76.70
06/13/2001 13:10	06/20/2001 10:59	165.8	43.80	86.60	79.16
06/20/2001 12:21	06/27/2001 14:23	170.0	50.48	84.32	77.40
06/27/2001 15:23	07/03/2001 11:34	140.2	31.43	83.61	70.23
07/03/2001 12:49	07/11/2001 15:53	195.1	30.86	86.74	70.56
07/11/2001 16:01	07/18/2001 15:36	167.6	34.41	87.73	70.31
07/18/2001 15:41	07/25/2001 11:41	164.0	54.32	82.76	72.04
07/25/2001 11:46	08/02/2001 9:13	189.5	46.92	84.89	71.82
08/02/2001 9:26	08/09/2001 9:01	167.6	44.65	88.73	73.77
08/09/2001 9:10	08/16/2001 9:59	168.8	32.99	88.45	75.29
08/16/2001 10:04	08/23/2001 10:45	168.7	59.30	87.73	76.81
08/23/2001 10:50	08/30/2001 9:21	166.5	57.59	84.04	74.60
08/30/2001 9:48	09/06/2001 9:11	167.4	30.86	83.33	71.95
09/06/2001 9:19	09/13/2001 9:33	168.2	38.11	89.44	76.76
09/13/2001 9:36	09/20/2001 10:05	168.5	37.97	85.46	76.20
09/20/2001 10:14	09/27/2001 9:38	167.4	44.51	86.31	77.58
09/27/2001 9:48	10/02/2001 10:03	120.2	61.43	83.75	76.12
10/02/2001 10:07	10/10/2001 10:59	192.9	41.38	85.17	76.97
10/10/2001 11:04	10/18/2001 9:22	190.3	38.11	88.02	77.03
10/18/2001 9:32	10/25/2001 10:41	169.2	66.69	84.75	77.07
10/25/2001 11:34	11/01/2001 8:07	164.5	60.29	85.32	77.36
11/01/2001 9:35	11/08/2001 9:12	167.6	68.40	86.03	77.69
11/08/2001 9:18	11/15/2001 9:09	167.9	37.68	83.89	77.15
11/15/2001 9:23	11/21/2001 14:39	149.3	41.66	84.46	76.75
11/21/2001 14:45	11/28/2001 15:58	169.2	62.57	85.89	76.05
11/28/2001 16:32	12/06/2001 9:16	184.7	39.53	86.60	76.93
12/06/2001 9:46	12/13/2001 9:39	167.9	33.27	87.02	77.14
12/13/2001 9:58	12/20/2001 9:38	167.7	36.54	85.32	77.12
12/20/2001 10:05	12/27/2001 9:42	167.6	38.68	82.76	75.67
12/27/2001 9:51	01/03/2002 16:06	174.2	66.97	82.05	75.42
01/03/2002 16:20	01/10/2002 9:39	161.3	69.25	82.62	75.87
01/10/2002 9:46	01/17/2002 9:55	168.1	68.96	82.05	76.11
01/17/2002 10:02	01/24/2002 10:29	168.4	65.98	84.75	75.66
01/24/2002 10:41	01/31/2002 13:02	170.3	68.82	82.76	76.75
01/31/2002 13:06	02/07/2002 9:54	164.8	69.82	83.47	76.22
02/07/2002 10:30	02/14/2002 11:01	168.5	46.78	85.17	76.83
02/14/2002 11:13	02/21/2002 9:40	166.5	37.26	84.18	77.01
02/21/2002 10:01	02/28/2002 9:41	167.7	64.70	83.04	76.81
02/28/2002 9:45	03/07/2002 10:18	168.6	38.96	83.18	76.94
03/07/2002 10:31	03/14/2002 8:33	166.0	40.53	86.03	77.19
03/14/2002 9:14	03/21/2002 10:03	168.8	65.27	83.89	76.68
03/21/2002 10:30	03/25/2002 16:16	101.8	68.11	83.61	76.76
03/25/2002 16:20	04/02/2002 13:50	189.5	67.68	85.89	76.92
04/02/2002 13:53	04/09/2002 13:43	167.8	65.55	85.32	76.79
04/09/2002 14:51	04/16/2002 13:12	166.4	39.39	85.60	76.69
04/16/2002 13:15	04/23/2002 13:16	168.0	64.84	84.75	76.61
04/23/2002 13:21	04/30/2002 13:53	168.5	63.85	83.89	75.95
04/30/2002 14:02	05/07/2002 14:07	168.1	68.54	83.47	76.01
05/07/2002 14:10	05/15/2002 13:05	190.9	34.13	84.18	76.10
05/15/2002 13:19	05/22/2002 13:59	168.7	41.66	83.89	75.43
05/22/2002 14:03	05/29/2002 13:13	167.2	37.40	84.18	75.93
05/29/2002 13:16	06/03/2002 13:41	120.4	37.97	84.75	75.78
		0.0			
		SUMMARY:	30.86	89.44	
			MIN	MAX	

Location No. = **07**

Starting Date & Time	Ending Date & Time	Elapsed Time (hours)	Minimum psi	Maximum psi	Average psi
03/16/2001 13:26	03/20/2001 7:21	89.9	43.37	95.41	76.85
03/23/2001 10:35	03/30/2001 11:37	169.0	44.08	93.14	75.27
03/30/2001 11:37	04/06/2001 12:33	168.9	43.65	95.41	76.15
04/06/2001 12:33	04/12/2001 8:10	139.6	44.65	94.13	75.77
04/12/2001 8:15	04/19/2001 5:44	165.5	42.09	95.13	75.68
04/19/2001 10:30	04/25/2001 6:23	139.9	46.36	94.99	77.90
04/25/2001 12:49	05/02/2001 13:01	168.2	45.64	95.70	76.04
05/02/2001 13:08	05/09/2001 11:09	166.0	43.65	93.42	77.40
05/09/2001 12:21	05/15/2001 10:34	142.2	39.53	93.99	77.90
05/15/2001 16:16	05/22/2001 13:28	165.2	48.92	93.99	78.67
05/22/2001 15:49	05/29/2001 14:47	167.0	41.81	94.99	74.65
05/29/2001 17:47	06/05/2001 14:58	165.2	38.39	105.22	76.53
06/05/2001 16:13	06/12/2001 13:54	165.7	45.22	94.42	74.42
06/12/2001 16:37	06/19/2001 12:05	163.5	48.20	96.27	77.31
06/19/2001 15:20	06/26/2001 14:47	167.4	46.21	90.44	74.86
06/26/2001 16:00	07/02/2001 14:49	142.8	44.36	87.59	72.09
07/02/2001 15:26	07/10/2001 16:24	193.0	62.14	94.70	83.97
07/10/2001 16:41	07/17/2001 15:42	167.0	47.64	94.70	81.86
07/17/2001 16:23	07/24/2001 15:05	166.7	54.75	97.40	82.81
07/24/2001 16:02	07/31/2001 15:58	167.9	54.03	96.69	79.75
07/31/2001 16:21	08/07/2001 16:09	167.8	50.76	94.13	80.83
08/07/2001 16:13	08/14/2001 16:37	168.4	53.18	93.42	81.45
08/14/2001 16:49	08/21/2001 16:27	167.6	49.20	97.69	87.45
08/21/2001 17:12	08/28/2001 16:34	167.4	61.57	93.99	87.87
08/28/2001 17:01	09/04/2001 16:50	167.8	49.48	89.87	75.55
09/04/2001 17:06	09/11/2001 10:19	161.2	53.47	93.99	87.06
09/11/2001 18:08	09/18/2001 16:48	166.7	53.47	93.99	89.15
09/18/2001 16:52	09/25/2001 16:34	167.7	77.78	97.97	91.86
09/25/2001 17:48	10/01/2001 15:21	141.5	61.57	93.99	84.72
10/01/2001 17:22	10/09/2001 9:44	184.4	65.69	93.99	87.80
10/09/2001 17:02	10/16/2001 8:57	159.9	49.48	93.99	88.67
10/16/2001 16:45	10/24/2001 7:39	182.9	49.48	93.99	79.59
10/24/2001 10:38	10/31/2001 8:38	166.0	49.20	92.88	77.80
10/31/2001 16:30	11/07/2001 7:32	159.0	49.20	93.71	79.68
11/07/2001 16:30	11/14/2001 7:20	158.8	49.58	93.11	78.77
11/14/2001 14:24	11/20/2001 8:54	138.5	1.42	94.42	76.67
11/20/2001 10:22	11/27/2001 9:46	167.4	0.43	95.13	76.99
11/27/2001 9:52	12/04/2001 16:41	174.8	46.21	95.98	78.07
12/04/2001 16:48	12/11/2001 16:02	167.2	31.00	99.54	77.07
12/11/2001 16:24	12/19/2001 14:58	190.6	36.54	96.98	77.68
12/19/2001 15:03	12/26/2001 18:00	171.0	42.37	97.97	77.41
12/26/2001 18:27	01/02/2002 16:22	165.9	43.09	96.27	76.55
01/02/2002 16:33	01/09/2002 13:33	165.0	42.66	94.42	76.45
01/09/2002 14:13	01/15/2002 16:11	146.0	42.37	95.41	76.07
01/15/2002 16:15	01/22/2002 16:37	168.4	44.65	95.21	77.36
01/22/2002 16:41	01/29/2002 14:48	166.1	42.37	95.70	76.69
01/29/2002 16:21	02/05/2002 15:30	167.2	43.65	94.99	76.49
02/05/2002 15:35	02/12/2002 15:54	168.3	44.36	95.98	77.45
02/12/2002 16:01	02/19/2002 15:39	167.6	44.65	96.27	80.07
02/19/2002 15:57	02/26/2002 15:14	167.3	46.21	96.69	77.36
02/26/2002 16:24	03/05/2002 16:29	168.1	44.36	96.98	79.05
03/05/2002 16:34	03/12/2002 15:59	167.4	36.83	95.41	77.92
03/12/2002 16:07	03/19/2002 16:47	168.7	44.36	95.41	77.14
03/19/2002 17:08	03/25/2002 16:28	143.3	43.94	94.13	77.33
03/25/2002 16:36	04/02/2002 17:00	192.4	43.37	95.41	77.72
04/02/2002 17:04	04/09/2002 14:22	165.3	43.37	96.27	77.59
04/09/2002 16:42	04/16/2002 12:48	164.1	44.36	95.70	76.37
04/16/2002 13:01	04/23/2002 12:16	167.3	44.65	97.40	75.95
04/23/2002 13:06	04/30/2002 8:45	163.7	44.68	93.71	75.65
04/30/2002 14:26	05/07/2002 13:01	166.6	44.08	94.42	75.75
05/07/2002 13:32	05/15/2002 11:48	190.3	40.67	95.13	75.28
05/15/2002 13:03	05/22/2002 14:11	169.1	42.66	96.41	75.33
05/22/2002 14:15	05/29/2002 12:54	166.6	42.94	95.70	74.85
05/29/2002 13:02	06/03/2002 14:23	121.4	44.36	93.14	71.24
		0.0			
		SUMMARY:	31.00	105.22	
			MIN*	MAX	

* Exceptions noted below

* The 11/16/01 2:16 pm low pressure (1.42 psi) incident at site #7 occurred when one of the maintenance personnel was out there recalibrating the transmitters and the water was automatically shut off for 5 minutes.

* The 11/20/01 3:22pm, low pressure (0.43 psi) incident at site #7 occurred when one of the maintenance personnel was out there recalibrating the transmitters and the water was automatically shut off for 20 minutes.

APPENDIX H
CALCULATING INTRUSION IN A DISTRIBUTION SYSTEM

Intrusion into a pipeline will obey the relationship for flow through an orifice. We have chosen the relationship used for sprinkler flow where the intrusion rate for a particular location is given by:

$$Qi = Ki \sqrt{(Pex - P)}$$

where

Qi = intrusion flow rate (g.p.m. or L/s)
Ki = intrusion constant
Pex = exit pressure (psi or k.p.a.)
P = internal pressure (psi or k.p.a.)

The intrusion constant defined in this manner is the intrusion flow rate in g.p.m. (or L/s) occurring for a pressure differential of 1 psi (or 1 k.p.a.).

The intrusion volume which occurs over a period of time is:

$$V_i = K_i \int_o^t \sqrt{(P_{ex} - P)}\ _{dt} = K_i\ I_i$$

where Ii is the intrusion integral.

The intrusion integral may be calculated for a transient event at any junction node in a distribution system where the internal pressure drops below the exit pressure. A feature has been added to Surge2000 to calculate the intrusion integral at all junction nodes. Because of the convention of concentrating demands and calculating pressure variations at junction nodes, intrusion calculations are also concentrated at junction nodes.

CALCULATING THE INTRUSION CONSTANT

It is assumed that the intrusion will be directly related to the leakage in a distribution system. Therefore, an estimate of the leakage can be used to determine the Intrusion Constant which can then be used to estimate the intrusion associated with a transient event. We define the leakage factor, Cf, as the ratio of the leakage to the average demands.

$$Cf = \text{leakage/average demand}$$

For example, a leakage factor of 0.1 would result in a leakage of 10% of the average demand. If we assume that the leakage also obeys the orifice relationship, then the leakage rate at a node is:

$$Q\ell = K\ell \sqrt{(Pex - Ps)}$$

where

$Q\ell$ = leakage flow rate (g.p.m. or L/s)
$K\ell$ = leakage constant
Ps = steady state line pressure
Pex = external pressure

If we further assume that the leakage constant is the same for each node, we get:

$$DACf = K\ell \sum \sqrt{(Ps\text{-}Pex)}$$

where DA = average demand for system

We can readily calculate the summation using the initial steady state pressures and calculate the leakage constant. Because leakage and intrusion will basically use the same passage ways, it is reasonable to assume that the leakage and intrusion constants are the same value.

EXAMPLE CALCULATION FOR THE DAVENPORT MODEL

For the Davenport Model (1125 junction nodes)

Total Demand (DA) = 12510 g.p.m.
We assume a leakage factor (Cf) = 0.1
The leakage (DACf) = 1251 g.p.m.

This results in an average leak of 1.112 g.p.m. at each node. Based on the initial steady state conditions for the average demand situation, the $\sum\sqrt{(Ps\text{-}Pex)} = 9354$ for Davenport. This was determined using the initial steady state pressure (the average steady state pressure is 69 psi) and an exit pressure of 0.

Therefore:

$$K\ell = 1251/9354 = 0.134$$

We assume:

$Ki = K\ell = 0.134$

This result will predict an intrusion rate of 0.134 g.p.m. for a 1 psi pressure differential. We used Ki = 0.134 for the intrusion calculations presented in this report.

SENSITIVITY STUDIES

Three factors, which affect intrusion, were not known specifically for the surge events being modeled. These include:

1) Wave Speed
2) Demand Factor
3) External Head

A baseline surge event based on Event #3 (see Chapter 9) was used to perform sensitivity studies on these parameters. The pressure wave speed can vary significantly based on pipe material and wall thickness. Even very small amounts of air in the system can significantly effect the wave speed. Table H.1 shows the effect the wave speed has on the intrusion calculation. Table H.2 shows the effect of applying different global demand factors and Table H.3 shows the effect of the external head on intrusion calculations. For calculation of intrusion potential for case studies 1-14, a Wave Speed of 3000 feet/second, a Demand Factor of 1.0 and an External Head of 0 feet were used unless otherwise specified (Table H.4).

Table H.1
Effect of Wave Speed on Intrusion

Wave Speed (feet/second)	Number of Nodes	Total Time (seconds)	Intrusion Integral	Intrusion (gallons)
2000	70	222	7.1	0.96
2500	113	436	17.7	2.39
3000	132	541	22.6	3.05
3500	125	528	21.7	2.93
4000	113	483	19.6	2.64

Table H.2
Effect of Global Demand Factor (GDF) on Intrusion

GDF	Number of Nodes	Total Time (seconds)	Intrusion Integral	Intrusion (gallons)
0.5	140	601	26	3.51
1.0	132	541	22.6	3.05
1.5	118	476	19.4	2.62

Table H.3
Effect of External Head (Hex) on Intrusion

Hex (feet)	Number of Nodes	Total Time (seconds)	Intrusion Integral	Intrusion (gallons)
0	132	541	22.6	3.05
1	133	558	23.7	3.20
2	134	575	24.8	3.35
3	137	596	25.9	3.50
4	138	613	27.0	3.64
5	141	634	28.2	3.81

Table H.4
Intrusion Calculations for Model Simulations 1-14

Case	Number of Nodes	Total Time (seconds)	Intrusion Integral	Intrusion (gallons)
1	132	541	22.6	3.05
1a	118	476	19.4	2.62
1b	140	601	26	3.51
2	154	1041	39	5.27
3	108	381	15.2	2.05
4	0	0	0	0
5	124	503	2	2.89
6	0	0	0	0
7	0	0	0	0
8	0	0	0	0
9	32	392	17.9	2.42
10	1	7	0.1	0.01
11	116	377	13.6	1.83
12	19	197	7.3	0.99
13	1	3	0.1	0.01
14	19	196	0.1	1

REFERENCES

Abbaszadegan, M., P. Stewart, and M. LeChevallier. 1999. A strategy for the detection of viruses in groundwater by PCR. *Appl. Environ. Microbiol.* 65(2): 444-449.

Aboytes, R. and M. W. LeChevallier. 2002. Detection of Infectious *Cryptosporidium* in Filtered Drinking Water. Report to the American Water Works Service Company, Inc., Voorhees, NJ.

American Public Health Association, American Water Works Association, Water Environment Federation. 1998. Standard Methods For The Examination Of Water And Wastewater, 20th ed. Washington, DC: American Public Health Association.

American Water Works Association. 1992. AWWA C651 – AWWA Standard for Disinfecting Water Mains. Denver, CO: AWWA.

Armon, R. A. and P. Payment. 1987. A Modified m-CP Medium for Enumerating *Clostridium perfringens* from Water Samples. *Can. J. Microbiol.,* 34: 78-79.

Barber, L. B. and J. H. Writer. 1998. Impact of the 1993 Flood on the Distribution of Organic Contaminants in Bed Sediments of the Upper Mississippi River. *Environmental Science Technology* 32(14):2077-2083.

Barwick, R. S., D. A. Levy, G. F. Craun, M. J. Beach, and R. L. Calderon. 2000. Surveillance for waterborne disease outbreaks -- United States 1997-1998. *MMWR* 49(SS04); 1-35. Accessed at: http://www.cdc.gov/mmwr/preview/mmwrhtml/ss4904a1.htm.

Bennett, J. V., S. D. Holmberg, M. F. Rogers, and S. L. Soloman. 1987. Infectious and Parasitic Diseases. In. R. W. Amler and H. B. Dull (eds.), *Closing the Gap: The Burden of Unnecessary Illness.* Oxford University Press, New York, NY.

Bukhari, Z., T.M. Hargy, J.R. Bolton, B. Dussert, and J.L. Clancy. 1999. Inactivation of *Cryptosporidium parvum* Oocysts Using Medium-pressure Ultraviolet Light. *Jour. AWWA*, 91(3):86-94.

Bukhari, Z., M.M. Marshall, D.G. Korich, C.R. Fricker, H.V. Smith, J. Rosen, and J. L. Clancy, 2000. Comparison of *Cryptosporidium parvum* viability and infectivity assays following ozone treatment of oocysts. *Appl. Environ. Microbiol.,* 66(7):2972-2980.

Chappell, C.L., P.C. Okhuysen, C.R. Sterling, and H.L. Dupont. 1996. *Cryptosporidium parvum*: Intensity and Infection and Oocyst Excretion Patterns in Healthy Volunteers. *J. Infect. Disease,* 173:232-236.

Clancy, J.L., 1998. Inactivation of *Cryptosporidium parvum* Oocysts in Water Using Ultraviolet Light. *Jour. AWWA,* 90(9):92-102.

Colford, J.M., J.R. Rees, T.J. Wade, A. Khalakdina, J. F. Hilton, I. J. Ergas, S. Burns, A, Benker, C. Ma, C. Bowen, D.C. Mills, D.J. Vugia, D.D. Juranek, and D.A. Levy. 2002. Participant blinding and gastrointestinal illness in a randomized, control trial of an in-home drinking water intervention. Emerging Infectious Diseases, 8(1): 29-36.

DiGiovanni, G.D., F.H. Hasemi, N.J. Shaw, F.A. Abrams, M.W. LeChevallier, and M. Abbaszadegan. 1999. Detection of Infectious Cryptosporidium parvum Oocysts in Surface and Filter Backwash Water Samples by Immunomagnetic Separation and Integrated Cell Culture-PCR. *Appl. Environ. Microbiol.* 65(8): 3427-3432.

DuPont, H. L., C. L. Chappell, C. R. Sterling, P. C. Okhuysen, J. B. Rose, and W. Jakubowski. 1995. The Infectivity of *Cryptosporidium parvum* in Healthy Volunteers. *New England Journal of Medicine*, 332:855-859.

Eisenberg, J. N. S., T. J. Wade, C. M. Vu, A. Hubbard, C. C. Wright, D. Levy, P. Jensen, and J. M. Colford. 2002. Risk factors in HOV-associated diarrhoeal disease: the role of drinking water, medication and immune status. *Epidemiol. Infect.* 128: 73-81.

Finch, G.R., C.W. Daniels, E. K. Black, F.W. Schaefer III, and M. Belosovic. 1993a. Dose response of *Cryptosporidium parvum* in outbred neonatal CD-1 mice. *Appl. Environ. Microbiol.,* 59:3661-3665.

Finch, G.R., E.K. Black, L. Gyurek, and M. Belosevic. 1993b. Ozone Inactivation of *Cryptosporidium parvum* in Demand-Free Phosphate Buffer Determined by *In Vitro* Excystation and Animal Infectivity. *Appl. Environ. Microbiol.,* 59:4203-4210.

Geldreich, E.E. and M. W. LeChevallier. 1999. Microbial Water Quality in Distribution Systems. pp. 18.1-18.49, In: R. D. Letterman (ed), *Water Quality and Treatment, 5th edition,* McGraw-Hill, NY.

Hellard, M.E., M. I. Sinclair, A. B. Forbes, and C. K. Fairley. 2001. A Randomized, Blinded, Controlled Trial Investigating the Gastrointestinal Health Effects of Drinking Water Quality. Environ Health Perspect 109:773–778.

Hoff, J. C. 1986. Inactivation of Microbial Agents by Chemical Disinfectants. Environmental Protection Agency EPA/600/S2-86/067, Cincinnati, OH.

Khramtsov, N. V., M. Tilley, D. S. Blunt, B. A. Montelone, and S. J. Upton. 1995. Cloning and Analysis of a *Cryptosporidium parvum* Gene Encoding a Protein with Homology to Cytoplasmic Form Hsp70. *J Eukaryot Microbiol.*, 42:416-22.

Kirmeyer, G. J., M. Friedman, K. Martel, D. Howie, M. LeChevallier, M. Abbaszadegan, M. Karim, J. Funk, and J. Harbour. 2001. *Pathogen Intrusion into the Distribution System.* #90835. Awwa Research Foundation and American Water Works Association. Denver, CO.

Korich, D.G., J.R. Mead, M.S. Madore, N.A. Sinclair, and C.R. Sterling. 1990. Effects of Ozone, Chlorine Dioxide, Chlorine, and Monochloramine on *Cryptosporidium parvum* Oocyst Viability. *Appl. Environ. Microbiol.,* 56:1423-1428.

LeChevallier, M. W. 1990. Coliform regrowth in drinking water: a review. *J. Amer. Water Works Assoc.* 82(11): 74-86.

LeChevallier, M. W., G. Di Giovanni, J. L. Clancy, Z. Bukhari, S. Bukhari, J. S. Rosen, J. Sobrinho, and M. M. Frey. 2002. *Source Water Assessment: Variability of pathogen Concentrations.* #90906 Awwa Research Foundation and American Water Works Association. Denver, CO.

LeChevallier, M. W., R. W. Gullick, M. R. Karim, M. Friedman and J. E. Funk. 2003. The potential for health risks from intrusion of contaminants into the distribution system from pressure transients. *J. Water and Health* 1: in press.

Morris, R. D. and R. Levine. 1995. Estimating the incidence of waterborne infectious disease related to drinking water in the United States. *Assessing and Managing Health Risk from Drinking Water Contamination: Approaches and Application, Proceedings of the Rome Symposium, September 1994.* IAHS 233: 75-88.

Okhuysen, P.C., C.L. Chappell, C.R. Sterling, W. Jakubowski, and H.L. Dupont. 1998. Susceptibility and Serologic Response of Healthy Adults to Reinfection with *Cryptosporidium parvum*. *Infection and Immunity,* 66:441-443.

Ortega, Y.R., R.R. Sheehy, V.A. Ama, K.K. Oishi, and C.R. Sterling. 1991. Restriction Fragment Length Polymorphism Analysis of *Cryptosporidium parvum* Isolates of Bovine and Human Origin. *J. Protozoology,* 38:540-541.

Payment, P., L. Richardson, J. Siemiatycki, R. Dewar, M. Edwardes, and E. Franco. 1991. Randomized trial to evaluate the risk of gastrointestinal disease due to consumption of drinking water meeting microbiological standards. *Am. J. Pub. Health* 81(6): 703-708.

Payment, P., E. Franco, and J. Siemiatycki. 1993. Absence of relationship between health effects due to tap water consumption and drinking water quality parameters. Wat. Sci. Tech. 27(3-4): 137-143.

Payment, P., J. Siemiatycki, L. Richardson, G. Renaud, E. Franco, and M. Prevost. 1997. A prospective epidemiological study of gastrointestinal health effects due to the consumption of drinking water. *Intern. J. Environ. Health Res.* 7:5-31.

Pontius, F. W. 1997. SDWA Then and Now. *JAWWA* 89(3): 11-36.

Reasoner D. J. and E. E. Geldreich. A new medium for the enumeration and subculture of bacteria from potable water. *Appl. Environ. Microbiol.* 49(1): 1-7, 1985.

Spinner, M. L., and G. D. DiGiovanni. 2001. Detection and identification of mammalian reoviruses in surface water by combined cell culture and reverse transcription-PCR. *Appl. Environ. Microbiol.* 67: 3016-3020.

Sobsey, M. D., and J. S. Glass. 1984. Influence of water Quality on Enteric Virus Concentration by Microporous Filter Methods. *Appl. Enviro. Microbiol.* 47: 956-960.

United States Geological Survey. 1996. Contaminants in the Mississippi River, 1987-92.

Upper Mississippi River Basin Association. 1997. Upper Mississippi River Spill Response Plan and Resource Manual.

US Environmental Protection Agency. 1989a. Drinking water; national primary drinking water regulations; filtration, disinfection; turbidity, *Giardia lamblia*, viruses, *Legionella*, and heterotrophic bacteria; final rule. *Federal Register* 54:27486--541.

US Environmental Protection Agency. 1989b. Drinking water; national primary drinking water regulations; total coliforms (including fecal coliforms and *E. coli*); final rule. *Federal Register* 1989;54:27544--68.

US Environmental Protection Agency. 1990. Drinking water; national primary drinking water regulations; total coliforms; corrections and technical amendments; final rule. Federal Register 55:25064--5.

US Environmental Protection Agency. 1991. Drinking water regulations: maximum contaminant level goals and national primary drinking water regulations for lead and copper; final rule. *Federal Register* 56:26460--4.

US Environmental Protection Agency. 1996a. National primary drinking water regulations: monitoring requirements for public drinking water supplies; final rule. *Federal Register* 61:24353--88.

US Environmental Protection Agency. 1996b. ICR Microbiology Laboratory Manual. EPA/600/R-95/178, April. Office of Research and Development, Washington, DC. http://www.epa.gov/nerlcwww/icrm1_6.pdf

U.S. Environmental Protection Agency. 1998a. SURF YOUR WATERSHED Database, (http://www.epa.gov/surf).

U.S. Environmental Protection Agency. 1998b. Better Assessment Science Integrating Point and Non Point Sources (BASINS) Database, Version 2.0, (http://www.epa.gov/OST/BASINS).

US Environmental Protection Agency. 1998c. National primary drinking water regulations: interim enhanced surface water treatment; final rule. *Federal Register* 63:69477--521.

US Environmental Protection Agency. 1998d. Announcement of the drinking water contaminant candidate list; notice. *Federal Register* 63:10274--87.

US Environmental Protection Agency. 1999. Method 1623: *Cryptosporidium* and *Giardia* in Water by Filtration/IMS/FA. EPA-821-R-99-006. U.S. Environmental Protection Agency, Office of Water, Washington, DC.

US Environmental Protection Agency. 2000. Stage 2 microbial and disinfection byproduct Federal Advisory Committee agreement in principle. *Federal Register* 65(251): 83015-83024.

US Environmental Protection Agency. 2001a. EPA actions to safeguard the nation's drinking water supplies. Office of Water, October, 2001. Accessed at http://www.epa.gov/safewater/security/secf.html.

US Environmental Protection Agency. 2001b. Method 1601: Male-specific (F+) and Somatic Coliphage in Water by Two-step Enrichment Procedure. January 2001, Draft. Office of Water. Washington, D.C. 20460.

USEPA. 2001c. Method 1602: Male-specific (F+) and Somatic Coliphage in Water by Single Agar Layer (SAL) Procedure. January 2001, Draft. Office of Water. Washington, D.C. 20460.

Volk, C., N. Welch, and M. LeChevallier. 1996. Survey of Assimilable Organic Carbon and Coliform Regrowth in the American Water System. Report to the American Water Works Service Company, Inc., Voorhees, NJ.

Volk, C., N. Welch, and M. LeChevallier.. 1997. Biodegradable Organic Matter: Significance and Changes During Conventional Water Treatment. Report to the American Water Works Service Company, Inc., Voorhees, NJ.

Wade, T. J., S. Sandhu, C. Wright, S. Lee, L. Kats, S. Shaw, M. LeChevallier, D. Levy, J.M. Colford, Jr. 2002. An increased in gastrointestinal illness associates with historic flooding of the Mississippi River: disease surveillance from a randomized trial of in-home drinking water treatment. Proceedings of the International Conference on Emerging Infectious Diseases, March 24-27, Atlanta, GA.

Woods, K. M., M. V. Nesterenko, and S. J. Upton. 1995. Development of a Microtitre ELISA to Quantify Development of *Cryptosporidium parvum In Vitro*. *FEMS Microbiol Lett.,* 128:89-94.

ABBREVIATIONS AND ACRONYMS

AHBSS	Acidified Hanks balanced salt solution
ATCC	American Type Culture Collection
AW	American Water
AWS	American Water System
AWWA	American Water Works Association
AwwaRF	Awwa Research Foundation
BDL	Below detection limit
CC-PCR	Cell culture polymerase chain reaction technique
CC-QSD	Quantitative cell culture PCR
CEC	Clancy Environmental Consultants
CFU	Colony forming units
CSO	Combined sewer overflow
DAPI	4',6-diamidino-2-phenylindolr
DBP	Disinfection by-product
ddH_2O	Double distilled water
DI	Distilled water
DIC	Differential interference contrast microscopy
DNA	Deoxyribonucleic Acid
dNTP	Deoxynucleotide Triphosphate
ERD	Entity Relationship Diagram
ESR	Equivalent Solids Reservoir
ESWTR	Enhanced Surface Water Treatment Rule
FBS	Fetal bovine serum
FITC-mAb	Fluorescein isothiocyanate labeled monoclonal antibody
FS	Flowing stream
g	Gravity Force
GWDR	Ground Water Disinfection Rule
HBSS	Hanks balanced salt solution
HCT-8	Human ileocecal adenocarcinoma cells
HEPES buffer	(N-[2-hydroxyethyl] piperazine-N□-[2-ethanesulfonic acid])
HUC	Hydrological Unit Catalog
IA	Iowa
ICR	Information Collection Rule
IFA	Immunofluorescence assay
IMS	Immunomagnetic separation

LT2ESWTR	Long Term 2 Enhanced Surface Water Treatment Rule
µL	Microliter
µg	Microgram
MCP	Magnetic particle concentrator
MPN	Most probable number
NTU	Nephelometric turbidity unit
PAC	Powdered activated carbon
PBS	Phosphate Buffered Saline
PCR	Polymerase Chain Reaction
PFU	Plaque Forming Unit
pH	Negative log of Hydrogen ion concentration - measure of acidity or basicity
POTW	Publicly owned treated water
QA/QC	Quality assurance / Quality control
QSD	Quantitative sequence detection
RL	Reservoir or Lake
SDS	Sodium Dodecyl Sulfate
SSC	Sodium Chloride / Sodium Citrate Buffer
STP	Sewage Treatment Plant
SUVA	Specific Ultraviolet Absorbance
TOC	Total Organic Carbon
TPMC	Technology, Planning and Management Corporation
U	International units
UMR	Upper Mississippi River
UPGMA	Unweighted pair group method using arithmetic averages
USCE	United States Corps of Engineers
USEPA	United States Environmental Protection Agency
USGS	United States Geological Survey
UV_{254}	Ultraviolet light of 254 nanometer wavelength
WSL	Wisconsin State Laboratory of Hygiene
WTP	Water treatment plant
WWTP	Wastewater treatment plant

IWA Publishing's authorised EU representative for General Product Safety Regulations is Diane D'Arras, 15 rue Duret, 75116 Paris, France, e-mail: safety@iwap.co.uk.

Printed and bound by CPI Group (UK) Ltd, Croydon, CR0 4YY
17/09/2026
02218799-0004